ELECTRON-STREAM INTERACTION WITH PLASMAS

ELECTRON-STREAM INTERACTION WITH PLASMAS

RICHARD J. BRIGGS

RESEARCH MONOGRAPH NO. 29
THE M.I.T. PRESS, CAMBRIDGE, MASSACHUSETTS

ACKNOWLEDGMENT

This is Special Technical Report Number 10 of the Research Laboratory of Electronics of the Massachusetts Institute of Technology.

The Research Laboratory of Electronics is an interdepartmental laboratory in which faculty members and graduate students from numerous academic departments conduct research.

The research reported in this document was made possible in part by support extended the Massachusetts Institute of Technology, Research Laboratory of Electronics, jointly by the U.S. Army (Electronics Materiel Agency), the U.S. Navy (Office of Naval Research), and the U.S. Air Force (Office of Scientific Research) under Contract DA36-039-AMC-03200(E); and in part by Grant DA-SIG-36-039-61-G14; additional support was received from the National Science Foundation under Grant G-24073.

PHYSICS

Copyright © 1964

by

The Massachusetts Institute of Technology

Library of Congress Catalog Card Number: 64-8719

Printed in the United States of America

FOREWORD

There has long been a need in science and engineering for systematic publication of research studies larger in scope than a journal article but less ambitious than a finished book. Much valuable work of this kind is now published only in a semiprivate way, perhaps as a laboratory report, and so may not find its proper place in the literature of the field. The present contribution is the twenty-ninth of the M.I.T. Press Research Monographs, which we hope will make selected timely and important research studies readily accessible to libraries and to the independent worker.

J. A. Stratton

PREFACE

In recent years, plasma physicists and electrical engineers have devoted a rapidly increasing amount of attention to the study of the instabilities caused by the streaming of charged-particle beams through a plasma. In some experimental situations, these instabilities are undesirable and pains are taken to eliminate them, whereas other workers have used the beam-plasma interaction as a mechanism for generating and heating the plasma.* The present theoretical study is concerned primarily with the latter type of situation, and the assumptions are made accordingly. However, some of the results, and many of the techniques and viewpoints, should be of wider applicability.

The approach presented in this monograph has been strongly influenced by the viewpoints that have been found useful in the analysis of the interaction of electron beams with circuits, which has primary application in the field of microwave beam tubes. The most prominent example of this influence is the devotion of an entire chapter to a general discussion of amplifying and evanescent waves, convective and nonconvective instabilities, and mathematical methods for distinguishing them. It is well known in the microwave tube field that the "instability" in a traveling-wave amplifier ("spatial" growth) is basically different from the "instability" in a backward-wave oscillator ("temporal" growth). In the case of beam-circuit interactions, the coupling is weak and the behavior is more intuitively obvious than in the case of beam-plasma interactions, where the "coupling" between the two systems (beam and plasma) can be strong. The chapter on absolute instabilities and amplifying waves provides a logical extension of these ideas on spatial and temporal growth to include cases of strong coupling.

The approach to the beam-plasma problem presented here is not the only possible one. It differs considerably, for example, from the approach adopted in the majority of papers in the Russian literature, where the velocity-distribution aspects of the problem are heavily emphasized while the finite dimensions of the system are largely ignored. We should also add that there have been few detailed comparisons between theory and experiment at the present time, and it is not clear whether many factors largely ignored here (and elsewhere), such as the finite gradients of the beam and plasma densities, do play an important role.

---

*L. D. Smullin and W. D. Getty, Phys. Rev. Letters, 9, 3 (1962); and J. Appl. Phys., 34, 3421 (1963).

   This monograph is derived from the author's thesis,* which
was supervised by Professor A. Bers. The active collaboration,
encouragement, and advice of Professor Bers on all aspects of
this work are greatly appreciated. The program of research on
beam-plasma interaction at M. I. T. is under the direction of Professor L. D. Smullin, who contributed many helpful ideas and
suggestions, and in particular suggested the problem of beam-
ion interaction which is considered here at some length. Professor H. A. Haus also provided many helpful comments, especially
on the research in Chapter 2.

   The author is indebted to all of his associates in the Research
Laboratory of Electronics for many discussions, and particularly
to Mr. Bruce Kusse and Mr. Satish Puri for the inclusion of some
of their S. M. thesis work in this monograph. Miss Susan Rosen-
baum of the Joint Computing Group of the Research Laboratory of
Electronics programmed most of the numerical calculations, which
were done at the Computation Center at M. I. T. The original man-
uscript and many additions were typed by Miss Ruth Fuller. My
wife, Kathleen, in addition to her innumerable other forms of support and assistance, expedited the preparation of the original man-
uscript by typing it in a rough draft form.

   This research was performed in the Research Laboratory of
Electronics of the Massachusetts Institute of Technology. The work
was supported in part by the U. S. Army, U. S. Navy, and U. S. Air
Force and in part by the National Science Foundation. Financial
assistance to the author for one year was provided by an R. C. A.
Industrial Fellowship from the Research Laboratory of Electronics.

Cambridge, Massachusetts
August 1, 1964   Richard J. Briggs

---

*Instabilities and Amplifying Waves in Beam-Plasma Systems,
Ph. D. Thesis, Department of Electrical Engineering, M. I. T.,
February, 1964.

CONTENTS

Chapter 1

INTRODUCTION

Recently there has been considerable interest in the problem of
the collective interactions of charged-particle beams (or streams)
with plasmas. Such interactions arise from a coupling of perturbations in the macroscopic density and current of the beam with
those in the plasma through the associated electromagnetic field.
Under certain conditions this coupling can lead to an increase in
the coherent oscillations of the beam and plasma particles, and in
the strength of the electromagnetic field, at the expense of the dc
energy of the beam; that is, these perturbations can be unstable.
These interactions, or instabilities, are of interest not only because of the important role that they play in some of the basic
physical processes occurring in plasmas but also from their possible applications as a means of amplifying microwave power and
as a means of heating a plasma.

The streaming of charged particles through a plasma arises
usually from either an externally injected beam of electrons (or
ions) or from currents induced in the plasma by external electromagnetic fields. The mathematical models adopted in this monograph are chosen largely with the first physical situation in mind,
namely, the case of an injected beam of electrons passing through
a plasma. In addition, a substantial portion of this investigation
is aimed at the application to plasma heating by the beam-plasma
interaction, and, more specifically, to the heating of the plasma
ions. For this reason, considerable attention is given to the low-
frequency interactions in beam-plasma systems with the plasma
electrons assumed to be relatively hot.

This work is concerned only with the linearized description of
the beam-plasma system, that is, with the "small-signal" perturbations on some unperturbed state. This approach allows one
to determine the conditions for which an instability occurs, and
also to say something about its initial rate of growth in time and
space, but it clearly does not provide any information on the resulting large-signal behavior. In all cases, the instabilities are
classified according to whether they are convective instabilities
(amplifying waves) that grow in space, or whether they are nonconvective instabilities (absolute instabilities) that grow in time,
when the system is uniform in (at least) one spatial dimension.
(These terms are defined more precisely in Chapter 2, where a

1

general mathematical procedure for distinguishing between convective and nonconvective instabilities is presented.) Moreover, an attempt is made, whenever possible, to determine the exact conditions for which a finite (nonuniform) system will become unstable.

It is, by now, well known that there is a rather large variety of different modes of instabilities in beam-plasma systems. In this monograph, a number of different limiting cases are analyzed in the hope of obtaining a more complete picture of the interactions. The interactions are analyzed in a one-dimensional system with a steady magnetic field aligned along the direction of the beam flow and in systems of finite size in the direction transverse to the beam velocity. The presence of the plasma ions is accounted for, and the interactions at both high and low frequencies are investigated. In every case, however, since a number of simplifying assumptions are necessary in order to make the analysis tractable, this study is by no means an exhaustive coverage of the problem of stream-plasma interaction.

There is a very close analogy between the interaction of an electron beam with a plasma and the beam-circuit interaction occurring in microwave beam tube amplifiers and oscillators. This analogy has been stressed by Smullin and Chorney[1,2] and Gould and Trivelpiece.[3] The analysis presented here has been rather strongly influenced by this point of view. In fact, the heavy emphasis placed here on classifying "instabilities" as amplification processes or "true" growth in time arises largely from the beam tube concept that the "instability" in a traveling-wave tube is quite a bit different physically from the "instability" in a backward-wave oscillator It will be shown that these beam tube analogies are very useful in the interpretation of the various instabilities, and, in fact, in some cases important parameters such as the critical length for oscillation can be obtained directly from the analogue.

1.1 Historical Review of the Problem

The concept of a macroscopic (collective) beam-plasma interaction was first proposed by Langmuir in 1925.[4] He proposed this interaction as a possible mechanism for generating the high frequency oscillations observed in his hot-cathode discharge. It took until the late 1940's, however, before widespread interest in the subject developed; this interest arose largely as a consequence of several important theoretical papers which were published during that time. Pierce,[5] in 1948, showed that a beam of electrons passing through a cold ion cloud should cause amplification of signals at frequencies just below the ion plasma frequency. He proposed this mechanism as a possible explanation for the spurious oscillations that were observed in some microwave tubes. Haeff,[6] in 1948 showed that amplification results when two electron beams move with different velocities, and he indicated that this could be a possible source of solar noise. Several authors, essentially simultaneously, proposed a two-stream microwave amplifier based on this principle.[7] Bailey[8] stated that the transverse as well as the longitudinal waves in beam-plasma systems in finite steady magnetic fields could be amplified; however, his interpretation of the transverse waves as "amplifying" was later criticized by Twiss.[9] Finally, Bohm and Gross[10] and Akhiezer and Fainberg[11] treated the electrostatic instabilities in the absence of a steady magnetic field, using the kinetic equations; the former authors gave a fairly complete description of the physical process of the energy transfer in the interaction.

It is, by now, well known that there are a number of different ways in which a beam of charged particles in the presence of a steady magnetic field can interact with a plasma, as will be discussed in more detail in Sections 1.1.1 and 1.1.2. Rather complete formulations of the unbounded beam-plasma system have been given which account for the velocity spreads in both the beam and the plasma, and some work has also been presented which accounts for the finite dimensions of the system transverse to the beam velocity. (Note that all of the earlier works[5-11] assumed an unbounded beam-plasma system.) The more recent references that are most pertinent to the work in this monograph are also briefly discussed. It should also be mentioned that rather extensive bibliographies to both the Russian and English literature have been given in two recent review articles.[12,13]

The following discussion is divided into sections dealing with theories for unbounded and bounded beam-plasma systems. The question of absolute instabilities and amplifying waves is briefly reviewed in the final section.

1.1.1 Unbounded Beam-Plasma Systems. The earliest work on beam-plasma interactions dealt only with the so-called electrostatic (or longitudinal space-charge-wave) instability in the absence of a steady magnetic field. In the simplest formulation of the problem, which neglects collisions and temperature, the spatial growth rate of this interaction is infinite at the electron plasma frequency, as was shown by Pierce[5] for the similar problem of a beam drifting through a cold ion cloud. This high-frequency instability has been investigated in some detail by a large number of authors; in particular, Sumi[14] and Boyd, Field, and Gould[15] showed that the amplification rate is bounded when the effects of collisions and temperature are included.

In a relatively cold plasma, the electrostatic instability is strongest near the electron plasma frequency. It has been shown by Rukhadze[16] and Kitsenko and Stepanov[17] that an electrostatic instability can occur at the ion plasma frequency when the unperturbed beam velocity is much less than the average thermal speed of the plasma electrons. This instability is clearly of great interest from the standpoint of heating the ions of a plasma by the interaction with an electron beam, and will be discussed at some length in this monograph.

Birdsall[18] has shown that the collisional damping in the plasma can itself be a mechanism for inducing an instability of the electrostatic wave. The idea that a lossy medium around an electron beam should result in amplification of the beam space-charge waves was first theoretically predicted by L. J. Chu[19] on the basis of his kinetic power theorem, and was later experimentally demonstrated by Birdsall, Brewer, and Haeff.[20]

It has been shown by a number of authors that the transverse waves in a magnetized plasma-beam system which propagate in the direction of the steady magnetic field can be unstable.[21-25] For an electron beam traversing a cold plasma, an instability of the transverse wave occurs near the ion cyclotron frequency. It has been pointed out by Stix[26] that the ion temperature of the plasma can be very important in this interaction. A detailed discussion of the effect of ion temperature on this interaction is given in Chapter 3.

1.1.2 Bounded Beam-Plasma Systems. Theories that account for the finite dimensions of the beam-plasma system in the direction transverse to the beam velocity usually deal with the cold, collisionless model of the plasma.

The space-charge-wave interaction of an electron beam with a cold, collisionless plasma in the presence of an infinite magnetic field in the direction of the beam velocity has been considered by Bogdanov, Kislov, and Tchernov[27] and Vlaardingerbroek, Weimer, and Nunnink.[28] The latter authors show that the amplification rate of this interaction is infinite only when the beam space-charge wavelength is (roughly) less than the transverse dimension of the plasma, for the case of both the beam and the plasma filling a waveguide.

The space-charge-wave interactions in the case of a finite axial magnetic field has been considered by several authors with the aid of the quasi-static approximation.[1,29-34] There have been, however, some difficulties in the interpretation of solutions of the dispersion equation in this case, as regards the meaning of the roots of complex wave numbers for real frequemcy in the vicinity of the negative dispersion wave in the plasma. This point is clarified in the present work by use of the amplification criteria developed in Chapter 2.

Smullin and Chorney[1,2,35] considered the interaction of an electron beam with an ion cloud within a quasi-static approximation and stressed the very close analogy of these results with the interactions in various types of microwave beam tubes. For the case of both the beam and the ion cloud filling a waveguide, they

showed that cyclotron-wave interactions could result as well as
the usual space-charge-wave interactions. Morse[36] and Getty[37]
used a similar model to analyze the interaction of an electron
beam with the plasma electrons and presented computations of
the dispersion near the interaction of the space-charge waves
and slow cyclotron wave of the beam with the negative-dispersion
plasma wave.

Kino and Gerchberg[38] have recently pointed out that a very thin
electron beam can have transverse as well as longitudinal modes
of instability when passing through a plasma of infinite extent.
These transverse modes of instability are obtained for zero external magnetic field as well as for finite magnetic fields.

Some work has also been started on analyzing the beam-plasma
interactions in a cold, collisionless plasma without making the
quasi-static approximation.[39,40] In addition, the effect of plasma
temperature in the case of beam-plasma systems of finite transverse dimensions has been included in a quasi-static formulation
by accounting only for the velocity spread in the direction parallel
to the external magnetic field.[41,42]

1.1.3 Absolute Instabilities and Amplifying Waves. The criterion of stability used by most authors is whether or not the dispersion equation of the uniform (infinite) system admits complex
values of the frequency for some real values of the wave number
(with the imaginary part of the frequency corresponding to growth
in time). It was first pointed out by Twiss,[43,44] and later by Landau and Liftshitz[45] and by Sturrock,[46] that two distinct types of instabilities can be identified physically. A spatial pulse on a uniform system can propagate along the system so that the disturbance
decays with time at a fixed point in space (convective instability)
or it can increase with time at every point in space (nonconvective
or absolute instability).

A somewhat related problem has arisen in the interpretation of
solutions of a dispersion equation for which complex values of the
wave number are found for real values of the frequency. This problem has arisen, for example, in the description of the spatial amplification process in the interaction of electron beams with circuits
in microwave beam tubes.[47] The solutions with complex wave numbers can represent amplifying waves which grow in space away
from some source, or they can represent evanescent waves which
decay away from the source. In simple cases, the concept of small-
signal energy and power has helped to resolve this question.[19,48-50]

In the course of this investigation, difficulties were encountered
in the interpretation of solutions of some of the dispersion equations by use of the existing mathematical criteria for distinguishing between absolute and convective instabilities, and between amplifying and evanescent waves.[46,51-53] New criteria that avoid these
difficulties are presented in Chapter 2. A critical review of the
previous criteria is presented at the end of that chapter.

1.2 Assumptions and Mathematical Models

As was mentioned before, the physical situation to which this theoretical investigation is intended to be most applicable is that of an externally generated electron beam that is injected into a plasma. For this reason, it will be assumed throughout that the beam is "cold," that is, that all beam electrons have the same unperturbed velocity. In many such experimental situations the beam density is usually several orders of magnitude less than the plasma density; for this reason it will often be useful to consider first the interactions in the mathematical limit of the beam density approaching zero. Interesting transitions can occur, however, for small but finite beam densities, and the majority of the analysis is by no means limited to the case of a beam of infinitesimal density.

The model of the plasma which is adopted is that of a fully ionized gas composed of particles which interact only through the large-scale, macroscopic electromagnetic fields. These particles can then be described within the framework of the collisionless Boltzmann-Vlasov equation. The effect of collisions on the beam-plasma interaction can safely be ignored if the frequency of the wave is much larger than any collision frequency, and means that the theory should be most applicable to hot plasmas of moderate or low density. It should also be mentioned that the effects of temperature are not included by means of a transport equation formalism, where only the first few moments of the Boltzmann equation are considered. Our analysis deals with the full distribution of velocities and includes the effects of the Landau and cyclotron damping.

The criteria for distinguishing between amplifying and evanescent waves and for determing absolute instabilities are presented in Chapter 2. These criteria are not restricted to the case of beam-plasma systems but are applicable to a wide class of uniform, time-invariant systems.

Chapter 3 considers the waves propagating along the steady magnetic field in an unbounded beam-plasma system. These waves consist of the electrostatic longitudinal wave, which is independent of the steady magnetic field, and the transverse circularly polarized waves in which the electromagnetic fields are perpendicular to the steady magnetic field. A detailed examination is made of the low-frequency longitudinal interactions for the case when the beam velocity is much less than the average thermal speed of the plasma electrons. The instabilities of the transverse waves occur at frequencies below the ion cyclotron frequency. The effects of the plasma temperature on these transverse instabilities are determined.

The analysis of the interaction in a cold plasma that is of finite extent in the plane transverse to the steady magnetic field is pre-

sented in Chapter 4. Various limiting cases are studied in order
to obtain analytical results. In Section 4.1 the case of a very low
density electron beam in a finite axial magnetic field is considered.
It is assumed that both the beam and the plasma fill a cylindrical
waveguide structure, and the quasi-static approximation is made.
The strength of the various interactions is determined for both
high and low frequencies. In Section 4.2 the interaction of a very
thin solid beam with a plasma that fills the waveguide is analyzed,
again within the quasi-static assumption. The interaction of solid
and hollow electron beams with a plasma in the presence of an
infinite axial magnetic field is analyzed in Section 4.3.

The low-frequency interaction with ions in a hot-electron plasma of finite transverse dimensions is considered in Chapter 5.
It is assumed that a large axial magnetic field confines the electrons to motion only along the field lines, so that only the random
velocities of the plasma electrons along the field lines are of importance. The beam and plasma are assumed to fill a cylindrical
waveguide structure, and the quasi-static assumption is made. It
is shown that the interaction in a hot-electron plasma of finite
transverse dimensions differs both qualitatively as well as quan-
titatively from the one-dimensional case.

Chapter 2

CRITERIA FOR IDENTIFYING AMPLIFYING WAVES AND ABSOLUTE INSTABILITIES

This chapter develops a general method for distinguishing between amplifying and evanescent waves and for detecting the presence of absolute instabilities. It is emphasized that although the subject of primary concern in this monograph is the beam-plasma interaction, the subject matter of the present chapter is not restricted to this special case. The approach presented here is a general one based only on an analysis of the dispersion equation and is therefore applicable to a wide class of uniform, time-invariant systems.

In Section 2.1 the problem is defined, and the instability terminology to be used throughout this monograph is explained. The mathematical proof of the method for identifying amplifying waves and absolute instabilities is given in Sections 2.2 and 2.3. The propagation of a pulse disturbance is studied in Section 2.4 to demonstrate the equivalence between "amplifying waves" and "convective instabilities." The reader interested primarily in the result will find the criteria restated in Section 2.5 along with some physical interpretations and comments on the mathematical procedure for applying the criteria. The discussion in Section 2.6 includes a comparison of the present formulation with the other work on this subject, some comments on the usefulness of such criteria, and a brief word about the concept of group velocity of propagating waves in unstable systems. Illustrative examples of the application of the criteria are given in Section 2.7; these examples include the simple quadratic equations obtained from the coupling-of-modes formalism.

2.1 Statement of the Problem

The general type of system considered in this chapter is time-invariant and uniform in (at least) one spatial dimension (the z coordinate). Because of this homogeneity in time and the spatial dimension z, linearized perturbations of the undriven system can be taken to be of the form $\exp[j(\omega t - kz)]$. The relation between the frequency ω and the wave number k is given by the dispersion equation

$$\Delta(\omega, k) = 0 \tag{2.1}$$

A problem of interest in plasma physics and elsewhere is that of determining the nature of "unstable waves" or "instabilities" in such a system. A wave is said to be unstable if for some <u>real</u> wave number k a complex $\omega = \omega_r + j\omega_i$ with negative ω_i is obtained from the dispersion equation, signifying growth in time of a spatially periodic disturbance of infinite extent. It was first indicated by Twiss[43,44] and Landau and Liftshitz,[45] and very clearly pointed out by Sturrock,[46] that two types of instabilities can be distinguished physically: "convective" instabilities, and "absolute" or "nonconvective" instabilities. In an infinite system, a pulse disturbance that is initially of <u>finite</u> spatial extent may grow in time without limit at <u>every</u> point in space* (an absolute instability) or it may "propagate along" the system so that its amplitude eventually decreases with time at any <u>fixed</u> point in space (a convective instability).†

It is perhaps at first surprising that a disturbance that is of finite spatial extent does not blow up in time in <u>every</u> case where the real wave numbers corresponding to unstable solutions of the dispersion equation are "excited." The representation of a spatially bounded disturbance, however, requires the superposition of many real wave numbers in the form[45,46]

$$\int_{-\infty}^{+\infty} f(k) e^{j[\omega(k)t - kz]} \frac{dk}{2\pi} \qquad (2.2)$$

(see also Appendix A). The limiting value of this integral as $t \to \infty$ (and z is held constant) is not necessarily infinite, even when some real k values yield solutions of the dispersion equation $[\omega(k)]$ with $\omega_i < 0$, since a decaying function can be represented as a superposition of many growing exponentials, as in the usual theory of Laplace transforms. The physical reason for this is that the pulse disturbance may convect away from its origin as it grows in amplitude, as was pointed out by Sturrock.[46]

---

*In a physical system, the amplitude of the oscillation is, of course, limited by nonlinear effects. For clarity in the discussions that follow, however, we will loosely refer to cases where the linearized analysis indicates exponential growth in time as a "response tending to infinity."

†Clearly, the labeling of an instability is always with respect to a particular reference frame, since a convective instability would appear as an absolute instability to an observer moving along with the "pulse." Thus one should realize that the term "absolute" instability adopted here does not imply growth in time at every point in space <u>in every reference frame.</u>

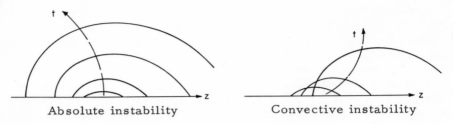

Absolute instability   Convective instability

Figure 2.1. Evolution of pulse disturbance in an unstable system.

To illustrate these ideas, "snapshot" views of some hypothetical convective and absolute instabilities are shown in Figure 2.1. We see that the distinguishing characteristic of an absolute instability is that it "spreads out" in both directions at once so that when the disturbance reaches a point, this disturbance keeps on growing in time at this point. The convective instability, on the other hand, "propagates along" the system as it grows in time so that the disturbance eventually disappears if one stands at a fixed point. Another physical interpretation of the distinction between these two is that the presence of an absolute instability implies that the system has an "internal feedback" mechanism so that oscillations can grow in time without the necessity of reflections from some termination of the system, whereas a convective instability requires such reflections (or an external feedback) for oscillations to grow exponentially in time at any fixed point in space.

In many cases one also may be interested in the sinusoidal steady-state response of a system at a particular (real) frequency. (It is crucial, however, as the development in this chapter shows, to ascertain whether such a steady-state in time can exist.) A problem which then arises is the interpretation of solutions that yield complex wave numbers $k = k_r + jk_i$ for this real frequency. In a "passive" system, as, for example, an empty waveguide, one would state on purely physical grounds that this solution represents decay in space away from some source; that is, it represents an "evanescent" (decaying) wave. In an "active" system that has a "pool" of energy in its unperturbed state, however (as, for example, a system containing an electron beam), such a solution could represent spatial growth of a sinusoidal time signal. That is, it could represent an "amplifying wave." In complicated cases, it is often not clear physically which situation prevails. One of the main purposes of this analysis is to determine a mathematical procedure for distinguishing such waves. Note that in the following, the terms "amplifying" and "evanescent" wave will be used only in connection with real values of the frequency ω. In addition, note that we are not restricting the term "evanescent wave" to the case of lossless systems, as is sometimes done.

In Sturrock's development,[46] he concludes that a convective instability is basically of the same type as an amplifying wave. That is, the process of spatial amplification of a sinusoidal time signal is really a form of "spatial instability" of the system. This result is also borne out by the analysis in the following sections, and the connection between the two is considered in some detail in Section 2.4.

Since in complicated cases there may be many solutions of the dispersion equation that have complex k for real ω, the criteria on amplifying and evanescent waves developed in the following sections are needed in order to sort out <u>which</u> imaginary part of complex k for real ω represents the "spatial growth rate" of a convective instability. In most cases, this growth rate in space at a <u>real</u> frequency is a more useful measure of the strength of a convective instability than is the maximum negative imaginary part of ω for real k (temporal growth rate).

2.2 Green's Function Formalism for the Response to a Localized Source

In order to determine the physical meaning of the roots of the dispersion equation as discussed in the previous section, we shall consider explicitly the excitation of these waves by a source. The simplest situation to investigate is that of a system which is infinitely long in the z-direction and excited by a source confined to a finite region of space (Figure 2.2). The response of the system outside of the source region $|z| > d$ is a linear combination of some of the "normal modes," or "natural responses," of the system. These normal modes are given by the solutions to the dispersion equation (2.1).[*]

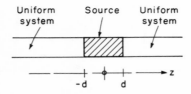

Figure 2.2. Driven system.

If we were considering only the question of distinguishing between amplifying and evanescent waves, we might be tempted to assume that sinusoidal steady-state conditions prevail, and pro-

---

[*]The term "normal modes" will be taken to mean both the solutions for k at some fixed ω from Equation 2.1, and the solutions for ω at some fixed k. Exactly which situation prevails in the following should be clear from the context.

ceed to investigate whether or not any waves can be excited which
grow in space away from the source region. There are at least
two objections to this procedure: (1) In order to neglect the reflections from terminations of the system, we must let the length
of the system approach infinity before (or faster than) we let time
approach infinity (to attain steady-state conditions). (2) We may
anticipate that if absolute instabilities are present, the system
will never attain such a steady state.

Both of these difficulties can be avoided by considering the excitation of this infinite system by a source that is zero for t < 0.
This allows us to study the manner in which the system approaches
the steady state, if indeed it does so at all. If we look at the asymptotic time response of the system at some fixed position outside of the source region, we may find that there is a disturbance
increasing exponentially with time, in which case there is an absolute instability. On the other hand, if there are no absolute instabilities, this asymptotic response, for the case of a sinusoidal
excitation, should be sinusoidal with time at the source frequency.
If the asymptotic time response contains any normal modes that
are spatially increasing away from the source region, these are
clearly amplifying waves.

In this approach, we are suppressing the role of any terminations of the system in order to establish this basic "causality" of
the waves on the uniform system; however, we should always keep
in mind that these terminations may play an important role in the
behavior of a given physical system. This is discussed more fully
in Section 2.6.

We will indicate the response of the system in Figure 2.2 by the
variable $\psi(t, z, \overline{r}_T)$, which symbolizes any (or all) of the physical
variables in the problem. Here $\overline{r}_T$ is the position vector in the
plane transverse to the z-direction. Similarly, the "source function" will be written as $s(t, z, \overline{r}_T)$. The response can be given in
terms of the source by a relation of the form[54]

$$\psi(t, z, \overline{r}_T) = \int K(t - t', z - z', \overline{r}_T, \overline{r}_T') s(t', z', \overline{r}_T') \, d^2\overline{r}_T' \, dz' \, dt'$$

$$(2.3)$$

where K is the Green's function; that is, it is the response at the
position $(t, z, \overline{r}_T)$ arising from an impulse source located at $(t', z', \overline{r}$
and the integration is over the "space-time volume" occupied by the
source s. Here, K is a function of $(t - t')$ and $(z - z')$, rather than
each of these variables separately, since the system is homogeneou
in these coordinates. For notational convenience, we will take the
source function to be of the form

$$s(t, z, \overline{r}_T) = T(\overline{r}_T) g(z) f(t)$$

$$(2.4)$$

where $f(t) = 0$ for $t < 0$. This form of the source function is sufficiently general for our purposes. We now perform Laplace
transformations with respect to time and a Fourier transformation with respect to the spatial coordinate z. A Fourier transform in space can always be performed for all finite times because of the finite speed of propagation of any disturbance. The
form of these transformations is illustrated now for the source
functions g(z) and f(t):

$$g(z) = \int_{-\infty}^{+\infty} g(k)e^{-jkz} \frac{dk}{2\pi} \tag{2.5}$$

$$g(k) = \int_{-\infty}^{+\infty} g(z)e^{jkz} \, dz \tag{2.6}$$

and

$$f(t) = \int_{-\infty-j\sigma}^{+\infty-j\sigma} f(\omega)e^{j\omega t} \frac{d\omega}{2\pi} \tag{2.7}$$

$$f(\omega) = \int_{0}^{\infty} f(t)e^{-j\omega t} \, dt \tag{2.8}$$

The integration in Equation 2.5 is carried out along the real-
k axis, and the integration in Equation 2.7 is carried out along
the line $\omega_i = -\sigma$. The integration in Equation 2.7 must be carried out <u>below</u> all singularities of $f(\omega)$ in order that $f(t)$ be zero
for $t < 0$. Similar transforms apply to all other quantities; these
transforms will always be written with the same symbol as the
physical variable except that the functional dependence is replaced
by ω and/or k.

For the purpose of distinguishing between amplifying and evanescent waves at some real frequency ω_0, we will usually consider
an excitation of the form

$$f(t) = e^{j\omega_0 t} \tag{2.9}$$

and therefore

$$f(\omega) = \frac{1}{j(\omega - \omega_0)} \qquad (2.10)$$

Also note that since we are assuming a localized source, $g(z) = 0$ for $|z| > d$; from this fact and from an inspection of Equation 2.6 it follows that $g(k)$ is an entire function of k (has no poles in the finite k-plane) as long as $g(z)$ is a "reasonable" function of z. As an example of $g(z)$, consider the spatial pulse shown in Figure 2.3. The transform in this case is just

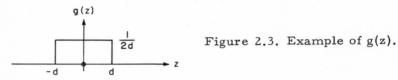

Figure 2.3. Example of $g(z)$.

$$g(k) = \frac{\sin (kd)}{kd} \qquad (2.11)$$

By applying the transforms to Equation 2.3, the transform of the response can be written as

$$\psi(\omega, k, \overline{r}_T) = G(\omega, k, \overline{r}_T)g(k)f(\omega) \qquad (2.12)$$

where

$$G(\omega, k, \overline{r}_T) = \int K(\omega, k, \overline{r}_T, \overline{r}_T')T(\overline{r}_T') \; d^2\overline{r}_T' \qquad (2.13)$$

The function $G(\omega, k, \overline{r}_T)$ is just the transform of the Green's function "weighted" by the transverse dependence of the source function. In simple cases, the source function $T(r_T')$ can be chosen so as to select only one of the transverse eigenmodes for consideration at a time, although it is not necessary that this be done. The actual response in space and time is recovered by applying inverse transforms; it can be written in the form

$$\psi(t, z) = \int_{-\infty}^{+\infty} \int_{-\infty-j\sigma}^{+\infty-j\sigma} G(\omega, k)f(\omega)g(k)e^{j(\omega t - kz)} \; \frac{d\omega \; dk}{(2\pi)^2} \qquad (2.14)$$

where the dependence on $\overline{r}_T$ is suppressed from here on for simplicity in notation.

Equation 2.14 summarizes the desired formalism; in the next section the general character of the asymptotic limit of the response in time will be determined by investigating this integral expression.

2.3 Proof of Criteria on Amplifying Waves and Absolute Instabilities

The general formalism expressing the response of an infinitely long system to a localized source that is "turned on" at $t = 0$ was developed in the last section. In this section, we shall specialize to the case of a sinusoidal source in order to bring out the appearance of amplifying waves most clearly. The response given by Equation 2.14 can be written in the form

$$\psi(t, z) = \int_{-\infty-j\sigma}^{+\infty-j\sigma} F(\omega, z) f(\omega) e^{j\omega t} \frac{d\omega}{2\pi} \tag{2.15}$$

where we define

$$F(\omega, z) = \int_{-\infty}^{+\infty} G(\omega, k) g(k) e^{-jkz} \frac{dk}{2\pi} \tag{2.16}$$

and where $f(\omega)$ is given by Equation 2.10.

The integral in Equation 2.15 is carried out along a line below the real-ω axis, as shown in Figure 2.4. The causality condition

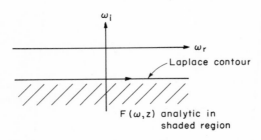

Figure 2.4. Analytic region of $F(\omega, z)$.

demands that $F(\omega, z)$ be analytic below the line $\omega_i = -\sigma$ in order that the response be zero for $t < 0$. A question which immediately arises is how large σ must be. The answer to this question will become clearer during the discussion of the analytic continuation of $F(\omega, z)$; however, one can predict in advance on purely physical grounds that it should sufficient for σ to be larger than the fastest growth rate in time of any unstable mode. That is, σ should be larger than the maximum negative imaginary part of ω for real k.

2.3.1 F(ω, z) as a Sum of Normal Modes. The function $F(\omega, z)$ contains the z-dependence of the response $\psi(t, z)$. Physically, we know that the response in a source-free region $|z| > d$ should be expressible as a sum over the normal modes of the undriven system. In simple cases, the Green's function $G(\omega, k)$ has poles in the complex k-plane (for some fixed complex ω on the Laplace contour) at just the "normal mode" wave numbers. These are the roots of the dispersion equation (2.1) for that particular ω. From Equation 2.6, and the fact that $g(z) = 0$ for $|z| > d$, it follows that $g(k)e^{-jkz} \to 0$ for $k \to -j\infty$ and $z > d$. As an example, consider the particular g(k) function given by Equation 2.11. The integral in Equation 2.16 can therefore be closed in the lower-half k-plane for $z > d$ as long as $G(\omega, k)$ is sufficiently well behaved at $k \to -j\infty$ (Figure 2.5). This assumption is a reasonable one, since this closure of the integral allows $F(\omega, z)$ to be expressed as a sum over the appropriate normal modes by the theory of residues. Note that g(k) is an entire function and therefore does not contribute any terms to the residue evaluation.

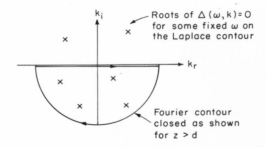

Figure 2.5. F(ω, z) as a sum of normal modes.

In more complicated cases, $G(\omega, k)$ for a fixed ω can have branch lines in the k-plane. This can be interpreted physically as a continuum of normal modes; as for example, the Van Kampen modes for longitudinal oscillations in a hot collisionless plasma,[55] or in cases involving radiation from open structures. For simplicity, we shall not consider these cases in the following discussion. This approach can be extended to cover these cases, but each case involving such branch lines must be handled individually. As an example, the branch lines that occur in the case of a hot, collisionless plasma are considered in Appendix B.

Having restricted ourselves to the case where the only singularities of $G(\omega, k)$ in the k-plane for some ω on the Laplace contour are poles at the roots of $\Delta(\omega, k) \lesseqgtr 0$, we can write the function $F(\omega, z)$ as a sum of normal modes in the form

$$F(\omega, z) = \sum - \frac{jg[k_{n+}(\omega)]}{\left[\frac{\partial}{\partial k} G^{-1}(\omega, k)\right]_{k=k_{n+}}} e^{-jk_{n+}z} \quad (2.17)$$

for $z > d$. The sum in Equation 2.17 is over all roots (k_{n+}) of $\Delta(\omega, k) = 0$ that have wave numbers k in the lower-half k-plane and where ω is some frequency on the Laplace contour $(\omega_i = -\sigma)$. For $z < -d$, the integral in Equation 2.16 is closed in the upper-half k-plane, and a similar expression for $F(\omega, z)$ valid for $z < -d$ is obtained, except that the sum is over all poles of $G(\omega, k)$ in the upper-half k-plane $[k_-(\omega)]$. Note that the dependence of the function $F(\omega, z)$ on z is that of (a sum of) exponential terms that all decay away from the source region for any ω on the Laplace contour.

2.3.2 Analytic Continuation of $F(\omega, z)$. The detailed response (for any given physical situation) could in principle be computed by carrying out the prescribed integration along the Laplace contour in Equation 2.15. Since our aim is only to discover some general characteristics of the asymptotic response, however, it is convenient to deform the Laplace integration as far as possible into the upper-half ω-plane. In this procedure, it is clear that the response in the limit as $t \to \infty$ is governed by the lowest singularity of $F(\omega, z)f(\omega)$ in the ω-plane. In particular, if $F(\omega, z)$ is analytic in the entire lower-half ω-plane and along the real-ω axis, the dominant term arises from the pole of $f(\omega)$ at ω_0, since the contribution from the rest of the integral becomes exponentially small as $t \to \infty$ (Figure 2.6).

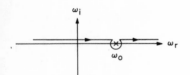

Figure 2.6. Deformed Laplace contour for asymptotic response of a system with no absolute instabilities.

From the preceding discussion, the singularities of $F(\omega, z)$ are clearly of prime importance in determining the asymptotic time response of the system. To explore the analyticity of $F(\omega, z)$, it is convenient to think of holding the real part of ω fixed while varying the imaginary part of ω (Figure 2.7a). Naturally, this process must be repeated for all real part of ω. From the original definition of $F(\omega, z)$, as given by Equation 2.16, (and from the representation in Figures 2.5 and 2.7), it is clear that $F(\omega, z)$

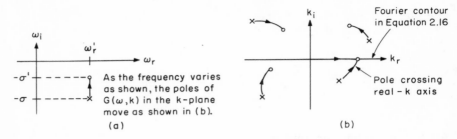

Figure 2.7. Exploring the analyticity of $F(\omega, z)$.

is a well-behaved function of ω unless one of the poles of $G(\omega, k)$ crosses the real-k axis, as is illustrated in Figure 2.7b. When this occurs, $F(\omega, z)$ as defined by Equation 2.16, jumps in value (as we cross the frequency $\omega = \omega_r^! - j\sigma^!$) by an amount equal to the residue at the pole that crossed the real-k axis. That is, in the ω-plane, the lines of complex ω for real k obtained by solving $\Delta(\omega, k) = 0$ for all real k are branch lines of the function $F(\omega, z)$ as defined by Equation 2.16 (Figure 2.8).

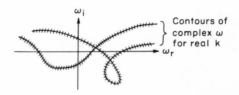

Figure 2.8. Branch lines of $F(\omega, z)$.

For a pole of $G(\omega, k)$ to cross the real-k axis for some ω in the lower-half ω-plane (such as $\omega^! = \omega_r^! - j\sigma^!$ in Figure 2.7), it must follow that the dispersion equation (2.1) yields complex ω solutions with negative imaginary parts for some real k. That is, it must be true that the infinite, homogeneous system supports unstable waves. It is now clear that σ should be chosen larger than the maximum growth rate in time of any unstable wave to satisfy the causality requirement, as was stated before. We assume that this maximum growth rate in time is bounded, that is, that there are no unstable waves with an infinitely fast growth rate in time.

The function $F(\omega, z)$ can be analytically continued through these branch lines; this analytic continuation is effected by redefining $F(\omega, z)$ in its integral form as

$$\widetilde{F}(\omega, z) = \int_C G(\omega, k)g(k)e^{-jkz} \frac{dk}{2\pi} \qquad (2.18)$$

where the contour C continues to include (or exclude) any poles
of $G(\omega, k)$ that cross the real-k axis as ω tends towards the real-ω
axis (Figure 2.9). The function $\widetilde{F}(\omega, z)$ is clearly the analytic continuation of $F(\omega, z)$, since they are identically equal below the line
$\omega_i = -\sigma$ and $\widetilde{F}(\omega, z)$ does not have these branch lines along the contours of complex ω for real k. The function $\widetilde{F}(\omega, z)$ can also be

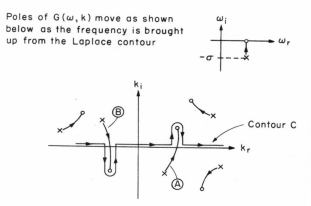

Figure 2.9. Definition of contour C.

given in the form of a sum over the poles of $G(\omega, k)$, as in Equation 2.17, with the implication that only the poles of $G(\omega, k)$ that
have k in the lower-half k-plane <u>when $\omega_i < -\sigma$</u> enter in the sum
(for z > d).

2.3.3 <u>Amplifying and Evanescent Waves.</u> If we assume for the
moment that this analytic continuation of $F(\omega, z)$ can be carried
out to a line just above the real-ω axis without encountering any
other singularities, then by deforming the integration over ω as
shown in Figure 2.6, we obtain the asymptotic dependence of the
response as

$$\lim_{t \to \infty} \psi(t, z) \to \widetilde{F}(\omega_0, z) e^{j\omega_0 t} \qquad (2.19)$$

from Equations 2.15 and 2.10.

In Figure 2.9 a pole such as A is clearly an amplifying wave,
since it has the space dependence exp $(+\alpha z)$ and it appears in the
response for z > d. Similarly, the pole B is an amplifying wave
that appears for z < -d. The other roots shown in the figure are
obviously evanescent waves. Thus, the determination of the locus
of k for all normal modes as ω is varied from ω_0 to $(\omega_0 - j\sigma)$ provides the desired criterion for distinguishing between amplifying
and evanescent waves.

2.3.4 Absolute Instabilities. There is a more fundamental difficulty with the analytic continuation of $F(\omega, z)$ when two poles of $G(\omega, k)$ merge <u>through</u> the contour C in the k-plane to form a double-order pole (Figure 2.10a). Since the integration in Equation 2.18

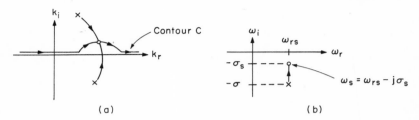

(a)   (b)

Figure 2.10. Merging of poles through contour C leading to an absolute instability.

must be carried out <u>between</u> the two merging poles, this results in a singularity of the function $\widetilde{F}(\omega, z)$ at that value of ω. In Figure 2.10b, as we let the "parameter" $\omega - \omega_s$ tend to zero, the two poles of k merge in the k-plane as shown. One would expect intuitively, without performing any detailed algebra, that the function $\widetilde{F}(\omega, z)$ should tend to infinity in the limit as $\omega - \omega_s$ tends to zero, since the integration path C becomes "stuck" and cannot be deformed around the two merging poles.[*]

The appearance of this singularity is perhaps best seen from the expression for $\widetilde{F}(\omega, z)$ as a sum of residues at the poles of $G(\omega, k)$, as given by Equation 2.17. At a double root of k, for some (ω_s, k_s), we have $(\partial/\partial k\ G^{-1})_{\omega_s, k_s} = 0$, and the dispersion equation near the double root is approximately

$$G^{-1}(\omega, k) \simeq \left(\frac{\partial G^{-1}}{\partial \omega}\right)_s (\omega - \omega_s) + \frac{1}{2}\left(\frac{\partial^2 G^{-1}}{\partial k^2}\right)_s (k - k_s)^2$$

$$(2.20)$$

Note also that the condition for a double root of k from the dispersion equation is the same as the condition for a saddle point of the function $\omega(k)$; that is, $\partial \omega/\partial k = 0$.

---

[*]One can easily check, by elementary means, that the integral along the real-x axis of $[1/(x - j\epsilon)]$ or $[1/(x - j\epsilon)^2]$ is <u>finite</u> in the limit as ϵ tends to zero, whereas the integral of $[1/(x - j\epsilon)(x + j\epsilon)]$ tends to infinity like $1/\epsilon$ as ϵ tends to zero.

If we use Equation 2.20 in Equation 2.17, we find that

$$\tilde{F}(\omega, z) \simeq \frac{g(k_s)e^{-jk_s z}}{\left[2\left(\frac{\partial G^{-1}}{\partial \omega}\right)\left(\frac{\partial^2 G^{-1}}{\partial k^2}\right)\right]^{1/2}_{\omega_s, k_s}} \frac{1}{(\omega - \omega_s)^{1/2}} \qquad (2.21)$$

near $\omega \simeq \omega_s$, <u>for either $z > d$ or $z < -d$</u>. (Equation 2.21 is correct within a $\pm$ sign, which can be determined only from a detailed consideration of the pole loci.) Therefore, this merging of the poles of $G(\omega, k)$ through the contour C leads to a branch pole of $\tilde{F}(\omega, z)$ at $\omega = \omega_s$. Note also that this branch pole is obtained for <u>both</u> $z > d$ and $z < -d$ and that the expression for $F(\omega, z)$ is the <u>same</u> in both regions for $\omega \simeq \omega_s$. This branch pole of $\tilde{F}(\omega, z)$ does <u>not</u> arise if two poles that are both below or both above the contour C merge into a double pole, since two terms (or none) then enter in the sum of residues in Equation 2.17, and these can be shown to cancel each other in the limit $\omega \to \omega_s$. This is to be expected, since a <u>double</u>-order pole that lies <u>inside</u> of a closed contour makes a finite contribution to the contour integral when the integral is evaluated by residue calculus. That is, the limit of $\tilde{F}(\omega, z)$, as $\omega - \omega_s$ tends to zero, is now finite because the integration path C does not lie <u>between</u> the merging poles.

This branch pole of $\tilde{F}(\omega, z)$ must be taken into account in the integration in the ω-plane, and the lowest singularity in the ω-plane becomes the dominant term as $t \to \infty$ (Figure 2.11). (Note again

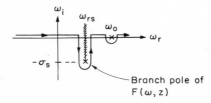

Figure 2.11. Integration in ω-plane with absolute instability.

that the <u>entire</u> lower-half ω-plane must be explored to determine which is the lowest such singularity.) In this case we have an absolute instability because the disturbance is blowing up in time at <u>every</u> point in space.[*] In the limit of $t \to \infty$, the asymptotic response can be evaluated as

---

[*]This formulation of the condition for an absolute instability has been given by Derfler[56] for the case of the double-stream instability in a plasma.

$$\psi(t, z) \simeq \frac{g(k_s)f(\omega_s)}{\left[2\left(\frac{\partial^2 G^{-1}}{\partial k^2}\right)\left(\frac{\partial G^{-1}}{\partial \omega}\right)\right]_{\omega_s, k_s}^{1/2}} e^{j(\omega_s t - k_s z)} \int \frac{e^{j(\omega - \omega_s)t}}{(\omega - \omega_s)^{1/2}} \frac{d\omega}{2\pi}$$

$$\simeq \frac{g(k_s)f(\omega_s)}{\left[2\pi j\left(\frac{\partial^2 G^{-1}}{\partial k^2}\right)\left(\frac{\partial G^{-1}}{\partial \omega}\right)\right]_{\omega_s, k_s}^{1/2}} \frac{e^{j(\omega_s t - k_s z)}}{t^{1/2}} \qquad (2.22)$$

where the double root of k occurs for $k = k_s$ and $\omega = \omega_{rs} - j\sigma_s$. It is interesting to note that, since k_s is in general complex, this asymptotic response can have an exponential envelope in space. The expression given in Equation 2.22 is not valid for very large z near the "wave front" of the disturbance because we have held z fixed as we let t approach infinity.

It is also possible that a triple pole of k can occur at some (ω_s, k_s) $[(\partial^2 G^{-1}/\partial k^2)_s = 0]$, or that $(\partial G^{-1}/\partial \omega)_s = 0$ at a double pole of k, and so forth. These singular cases will, in general, occur only for particular values of the system parameters, since they represent additional "constraints" at the point (ω_s, k_s). The extension of the present results to cover these singular cases has been given elsewhere.[57] In the general case, it is clear that a singularity of $\widetilde{F}(\omega, z)$ arises whenever one or more poles from one side of the C contour merge together with one or more poles from the other side of the C contour, regardless of the order of the root.

An entirely different type of singularity of $\widetilde{F}(\omega, z)$ can arise at a frequency ω_e for which $|k| \to \infty$, that is, at a frequency for which

$$\frac{1}{k^n} \sim C(\omega - \omega_e) \qquad (2.23)$$

as $\omega \to \omega_e$, where n is some integer (and, in fact, $n = 2$ in all physical examples of which the author is aware). This leads to an essential singularity of $\widetilde{F}(\omega, z)$ at $\omega = \omega_e$, since from Equation 2.17 we have

$$\widetilde{F}(\omega, z) \sim \exp\left[-jC^{-1/n} z(\omega - \omega_e)^{-1/n}\right] \qquad (2.24)$$

near $\omega \sim \omega_e$.[*] Note that this essential singularity of $\widetilde{F}(\omega, z)$ has nothing to do with poles of $G(\omega, k)$ merging through the C contour,

---

*There can be several terms of this form in the expression for $\widetilde{F}(\omega, z)$ if more than one root of k from Equation 2.23 lies on the appropriate side of the C contour. Also, if none of the roots of Equation 2.23 lies, for example, below the C contour, then $\widetilde{F}(\omega, z)$ will not have a singularity at ω_e for $z > d$.

and for this reason it can appear on one side of the source (say $z > d$) and not on the other ($z < -d$). (We recall that the branch pole type of singularity arising from merging poles, which was discussed before, must necessarily appear on <u>both</u> sides of the source.) If ω_e is in the lower-half ω-plane, it follows that in our model of the system we are allowing for unstable modes with an <u>infinitely short wavelength</u>. We can therefore ignore these essential singularities of $\tilde{F}(\omega, z)$ in the formulation of the stability criteria if we are analyzing a model in which there are no unstable modes with a finite growth rate in time for an infinitely short wavelength ($k \to \infty$). These essential singularities of $\tilde{F}(\omega, z)$ often <u>do</u> occur at <u>real</u> frequencies however, when one uses lossless models of the system, as, for example, a cold, collisionless model of a beam-plasma system. A simple example of this type is briefly discussed in Section 2.5.

2.3.5 <u>Response to a Pulse in Time</u>. The space-time dependence of the asymptotic response as given by Equation 2.22 is essentially independent of the source time function when there is an absolute instability present.* In particular, if $f(t)$ corresponds to a pulse in time, then $f(\omega)$ is an entire function of ω and the asymptotic response is always determined by the lowest singularity of $\tilde{F}(\omega, z)$ in the ω-plane, even if this singularity is in the upper-half ω-plane.

The consideration of a pulse excitation also verifies that the criterion derived is consistent with the physical description of absolute instabilities given in Section 2.1. For a pulse source, the response at the instant the source amplitude returns to zero will be of finite spatial extent, since the speed of propagation of all signals is finite. The behavior of this finitely extended waveform on the undriven system as $t \to \infty$ can then be determined; this response will remain finite or decay with time at every fixed point in space unless $\tilde{F}(\omega, z)$ has a branch pole in the lower-half ω-plane, that is, unless the system supports an absolute instability as determined by the criterion given in Section 2.3.4.

In the next section, a more detailed analysis of the propagation of pulse disturbances is given in order to bring out some additional aspects of the spatial and temporal growth of signals in "unstable" systems.

2.4 <u>Propagation of Pulse Disturbances and Relations Between Temporal and Spatial Growth Rates of Convective Instabilities.</u>

The usual statement on the "stability" of a wave is whether or not the dispersion equation yields complex ω for real k (with $\omega_i < 0$). It was stated in Section 2.1 that a pulse disturbance on

---

*This statement is true so long as the source does not have an exponentially increasing amplitude at a rate that is greater than the growth rate of the absolute instability.

a system which is "unstable" by this definition will always "blow
up" in amplitude, even though it may appear to decrease in time
at a fixed point because it could convect along the system as it
blows up. The validity of this statement is not obvious from the
analysis in Section 2.3, since we always investigated the asymptotic response at a fixed point in space. The statement is an important one, however, and is established in the present section.
It is shown, by allowing z and t to tend to infinity at a fixed ratio
(velocity), that a velocity can always be found for which a pulse
disturbance appears to increase exponentially with time at the
maximum growth rate of any unstable wave [maximum $(-\omega_i)$ for
real k]. This velocity is proposed as a definition of the "propagation velocity" of the pulse on an unstable system.

Sturrock, in his pioneering work,[46] noted a very close connection between the concepts of amplifying waves and convective instabilities. This connection is explored in the second part of this
section, where it is shown that the difference in these concepts
lies only in the excitation being considered (pulse or sinusoidal
in time) and not in the properties of the medium. It is shown, for
example, that in a system which supports convective instabilities
and has no absolute instabilities, there must exist one or more
roots of the dispersion equation with complex k for real ω which
are amplifying waves (that is, they enter in the response on the
side where they appear spatially growing and not decaying). In
addition, upper and lower bounds on the maximum amplification
rate in terms of the temporal growth rate of convective instabilities are given.

2.4.1 Propagation of a Pulse Disturbance. In the analysis of
Section 2.3, the response was investigated in the limit of $t \to \infty$
for fixed (finite) values of z. To demonstrate that a pulse disturbance does "blow up" in amplitude (even if it convects along
the system) in every case where the system supports unstable
waves, we shall investigate the impulse response of the system
in the limit of $t \to \infty$ and $z \to \infty$ with

$$z = Vt + z_0 \tag{2.25}$$

where V is a certain fixed velocity and z_0 remains finite. (We
could, of course, handle this problem by transforming into a
reference frame moving with respect to the laboratory frame.
We choose not to do this because of the complications introduced
by a relativistically correct transformation. The asymptotic
response we calculate, therefore, is the one measured in the laboratory frame with the laboratory time t, when the "measuring
instrument" moves with the velocity V.)

If the plot of complex ω for real k in the laboratory frame is
as shown in Figure 2.12, where σ_0 is the maximum negative im-

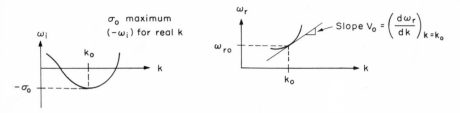

Figure 2.12. Sketch of complex ω for real k.

aginary part of ω for real k, then it will be proved that an observer moving with velocity

$$V_0 = \left(\frac{\partial \omega_r}{\partial k}\right)_{k=k_0}$$ (2.26)

will see the disturbance increase as exp $(\sigma_0 t)$. That is, the velocity given by Equation 2.26 is a sensible definition of the "propagation velocity" of the pulse on an unstable system, where k_0 is the real wave number for which the <u>maximum</u> negative imaginary part of ω for real k occurs, and ω_r is the <u>corresponding</u> real part of ω.

To prove this result, we write the impulse response $\psi(t, z)$ as a function of t and the <u>initial</u> position z_0, with $z(t)$ being given by Equation 2.25. From Equation 2.14, the impulse response can be written as

$$\psi(t, z_0) = \int_{-\infty-j\sigma}^{+\infty-j\sigma} \int_{-\infty}^{+\infty} G(\omega, k) e^{j(\omega - kV)t} e^{-jkz_0} \frac{dk\, d\omega}{(2\pi)^2}$$ (2.27)

For an impulse excitation in time and space, $s(\omega, k) = f(\omega)g(k) = 1$. If we define a new frequency variable

$$\omega' = \omega - kV$$ (2.28)

then we can write the response in a form which is completely analogous to that given in Equations 2.15 and 2.16:

$$\psi(t, z_0) = \int_{-\infty-j\sigma_0}^{+\infty-j\sigma_0} F'(\omega', z_0) e^{j\omega't} \frac{d\omega'}{2\pi}$$ (2.29)

where

$$F'(\omega', z_0) = \int_{-\infty}^{+\infty} G'(\omega', k)e^{-jkz_0}\frac{dk}{2\pi} \qquad (2.30)$$

and

$$G'(\omega', k) = G(\omega' + kV, k) \qquad (2.31)$$

The computation of the response as $t \to \infty$ now proceeds in exactly the same manner as that given in Section 2.3, except that ω is replaced by ω'.

We now note that if V is set equal to V_0 (Equation 2.26), the point $k = k_0$ and $\omega = \omega_{r_0} - j\sigma_0$ is a saddle point of $\omega'(k)$, a double root of k at that ω', because

$$\frac{d\omega'}{dk} = \frac{d\omega}{dk} - V_0 = \frac{d\omega_r}{dk} - V_0 + j\frac{d\omega_i}{dk} = 0 \qquad \text{at } k = k_0$$

$$(2.32)$$

In addition, it is clear from Equation 2.28 that Im ω' = Im ω for real k, and therefore the maximum negative imaginary part of $\omega'(k)$ for real k is also equal to σ_0. For this reason, we know that no poles of $G'(\omega', k)$ can have crossed the real-k axis for any Im $\omega' < -\sigma_0$; that is, the C contour defined in Section 2.3 is the real-k axis for Im $\omega' \leq -\sigma_0$ (Figure 2.13). The two contracting

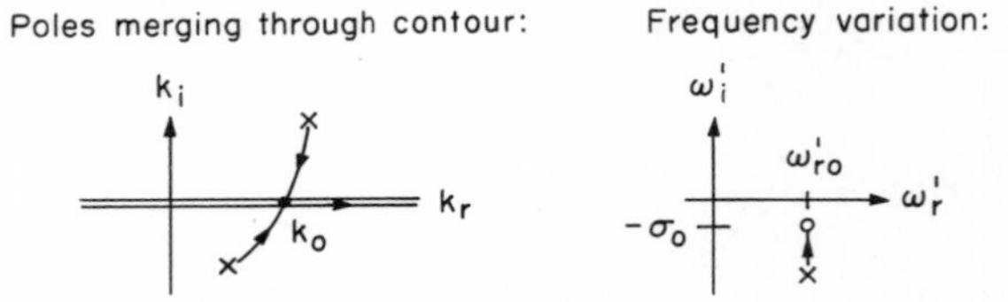

Poles merging through contour: Frequency variation:

Figure 2.13. Locus of poles in k-plane.

poles that form this saddle point of $\omega'(k)$ on the real-k axis (at $k = k_0$) merge as shown in Figure 2.13, and this saddle point of $\omega'(k)$ is a singularity of $F'(\omega', z_0)$ because the poles must be merging through the C contour. It then follows from the results of Section 2.3 that the response at any "initial position" z_0 increases in time as exp $(\sigma_0 t)$ for the choice $V = V_0$.

One might worry a bit about the special case where the merging poles shown in Figure 2.13 just "graze" the real-k axis. This situation would leave some doubt about whether or not the poles actually came from opposite halves of the k-plane for Im $\omega' < -\sigma_0$. The approach of ω to the saddle point, however, can be made at

any angle between 0° and -180°, and not just at -90° as shown in
the figure.

We also note, in passing, that it is quite obvious from the preceding formulation that no value of V can lead to an asymptotic
response which increases faster than exp $(\sigma_0 t)$.

2.4.2 Connection Between Amplifying Waves and Convective
Instabilities. The analysis in Section 2.3 indicates that there is
a very close connection between amplification and instability; an
amplifying wave must also, in a sense, be an unstable wave, since
the condition for the root to cross the real-k axis for some ω in
the lower-half ω-plane is precisely the same as the condition for
complex ω with a negative imaginary part for real k. It is clear,
therefore, that a necessary condition for a system to support amplifying waves is that complex ω with a negative imaginary part
be obtained from the dispersion equation. In the case of a system free from absolute instabilities, we might expect intuitively
that this condition should be sufficient as well; that is, it should
ensure the existence of amplifying waves for some real frequency.
This sufficiency is proved in the following.

We now specialize to the case of a system that has no absolute
instabilities and is driven by a source of the form

$$s(z, t) = \delta(z) f(t) \tag{2.33}$$

If we choose a sinusoidal source for f(t), as was done in Sec-
tion 2.3, the asymptotic response is given by $\widetilde{F}(\omega, z)$ with ω the
(real) frequency of the source (Relation 2.19). We can therefore
interpret $\widetilde{F}(\omega, z)$ as the "steady-state" response of the infinite
system to a sinusoidal drive.

If, on the other hand, we excite the system with an impulse
source f(t) = δ(t), then the impulse response ψ(t, z) can be given
in terms of this "steady-state" response as

$$\psi(t, z) = \int_{-\infty}^{+\infty} \widetilde{F}(\omega, z) e^{j\omega t} \frac{d\omega}{2\pi} \tag{2.34}$$

The integration over ω can be carried out along a line placed
an infinitesimal amount below the real axis because we are restricting our attention to systems free from absolute instabilities.
Equation 2.34 provides the desired connection between sinusoidal
and pulse responses, and hence also between amplifying waves
and convective instabilities.

In the first part of this section, it was shown that ψ(t, z) may
blow up in time even for convective instabilities if we take the
limit t → ∞ and z → ∞ with z and t related by Equation 2.25. In
the present formalism, the response in this same limit can be

written in the following form, where we express $\widetilde{F}(\omega, z)$ as a sum over the normal modes by Equation 2.17:

$$\psi(t, z_0) = \sum_{n} \int_{-\infty}^{+\infty} - \frac{je^{-jk_{n+}z_0}}{\left(\frac{\partial G^{-1}}{\partial k}\right)_{k=k_{n+}}} e^{j(\omega-k_{n+}(\omega)V)t} \frac{d\omega}{2\pi} \qquad (2.35)$$

We consider explicitly in the following only the case $V > 0$ and amplification in the +z direction; similar remarks apply to the case $V < 0$ and $z < 0$.

In the previous analysis, it was shown that $\psi(t, z_0)$ increases as $\exp(\sigma_0 t)$ as $t \to \infty$ for $V = V_0$, with V_0 given by Equation 2.26. It is clear by inspection of Equation 2.35, since ω is real in the integration, that one of the normal modes <u>must</u> be an amplifying wave over some band of real frequency for the integral in Equation 2.35 to diverge as $t \to \infty$. In fact, for $V = V_0$, the integral in Equation 2.35 will increase <u>more slowly</u> than $\exp(\alpha_M V_0 t)$, where α_M is the maximum amplification rate (maximum Im k_+ for real ω,

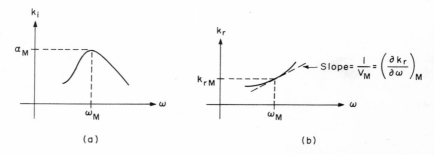

Figure 2.14. Sketch of complex k for real ω.

Figure 2.14a). We have therefore established the following lower bound on the maximum amplification rate (of a system free from absolute instabilities):

$$\alpha_M > \frac{\sigma_0}{V_0} = \frac{\max(-\omega_i) \text{ for real } k}{\frac{\partial \omega_r}{\partial k} \text{ at max}(-\omega_i)} \qquad (2.36)$$

The exact evaluation of the asymptotic limit of $\psi(t, z_0)$ from Equation 2.35 could be accomplished in principle by a saddle-point technique. In this method, the dominant contribution to the integral comes from integrating through the points of stationary phase wher

$$1 - \frac{\partial k(\omega)}{\partial \omega} V = 0 \tag{2.37}$$

Note that choosing the velocity V equal to V_M (Figure 2.14b) will make the point $k = k_{rM} + j\alpha_M$, $\omega = \omega_M$, a point of stationary phase, and hence for this velocity the response will increase as $\exp(\alpha_M V_M t)$ as $t \to \infty$. Since we showed in the first part of this section that for any velocity V the response will increase more slowly than $\exp(\sigma_0 t)$, we can also state the following <u>upper bound</u> on the maximum amplification rate:

$$\alpha_M < \frac{\sigma_0}{V_M} \tag{2.38}$$

In proving that this is an upper bound, however, we have made use of the fact that $\partial k_i / \partial \omega = 0$ at the maximum of $k_i(\omega)$ for real ω. Therefore, we have actually <u>assumed</u> in this proof that the amplification rate is less than infinity, since the zero derivative of $k_i(\omega)$ at the maximum would not be true for a case in which k_i has a pole at ω_M. Such cases are sometimes obtained when idealized models of the system are used.

2.5 Comments on the Application of the Criteria and Some Physical Interpretations

2.5.1 Amplifying Waves. The criteria on amplifying and evanescent waves developed in Sections 2.2 and 2.3 can be stated in the following manner: "To decide whether a given wave with a complex $k = k_r + jk_i$ for some real ω is amplifying or evanescent, determine whether or not k_i has a different sign when the frequency takes on a large negative imaginary part. If it does, then the wave is amplifying; otherwise, it is an evanescent wave."

This statement has a simple physical interpretation if we think of driving the system with a source that is localized in space and has an exponentially increasing sinusoidal time dependence.* For a sufficiently large exponential growth in time of the source, the principle of <u>causality</u> would imply that all waves should decay away from the source. Therefore, an amplifying wave should have the property that its "growth constant" <u>change sign</u> as the frequency acquires a sufficiently large negative imaginary part corresponding to exponential growth in time.

Normally, since one would like to know which waves are amplifying over the entire range of real frequency, the locus of the

---

*This interpretation was suggested by Professor A. Bers prior to the mathematical development given in Section 2.3, and was a substantial stimulus toward its development.

roots of k from $\Delta(\omega, k) = 0$ must be traced in the complex k-plane
for many such real frequencies. It is convenient, although not
necessary, to record the locus of these roots holding the real part
of ω fixed (Figure 2.15). Mathematically, this operation is just

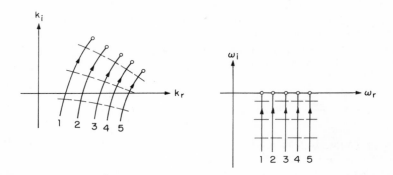

Figure 2.15. Mapping of $\omega(k)$ for an amplifying wave.

a mapping of the lines of constant real part of ω into the complex
k-plane; that is, our resulting locus plot is really a contour map
representation of the function $\omega(k)$ in the complex k-plane. For a
dispersion equation higher than first order in ω and k, of course,
there would be many sheets (or branches) of the function $\omega(k)$ being
traced, and not just one as depicted in Figure 2.15.

2.5.2 Absolute Instabilities. This mapping operation also indicates whether or not any absolute instabilities are present. An
absolute instability is obtained whenever there is a double root of
k for some complex ω in the lower-half ω-plane for which the two
merging roots come from different halves of the complex k-plane
(upper and lower) when ω has a large negative imaginary part (Fig
ure 2.16). The condition for a double root of k at $k = k_s$ for some
$\omega = \omega_s$ is the same as the condition that the function $\omega(k)$ have a
saddle point at $k = k_s$, since $(\omega - \omega_s) \sim (k - k_s)^2$ near this point.
The general form of such a saddle point should be as shown in
Figure 2.16. Its appearance should be obvious when the lines of
constant real part of ω are constructed in the k-plane even if the
"density" of these lines is fairly rough, as will be evident when
the criteria are applied to various examples in the later chapters.

There is an interesting feature about the saddle point of $\omega(k)$ illustrated in Figure 2.16 which also occurs in several physical examples discussed in Chapters 4 and 5. In the figure the root that
has complex k for real ω enters in the response for $z < -d$ when
$\omega_r < \omega_{rs}$ and enters for $z > d$ when $\omega_r > \omega_{rs}$ over the range of real
frequency being considered. Since $k_i > 0$ for real ω, this mode is,
in a sense, an "amplifying" wave for $\omega_r > \omega_{rs}$ and an "evanescent"
wave for $\omega_r < \omega_{rs}$. Since the asymptotic time response arises
largely from the contribution near the saddle point in the ω-plane

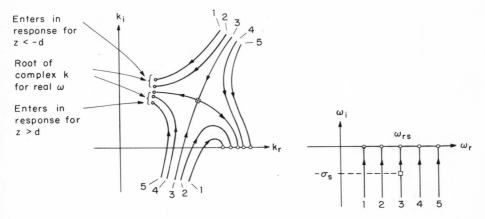

Enters in
response for
z < -d

Root of
complex k
for real ω

Enters in
response for
z > d

Figure 2.16. A saddle point of $\omega(k)$ indicating an absolute instability. Circles indicate real ω, square indicates saddle
point.

and not from the integral along the real ω-axis (see Figure 2.11),
this fact is of little consequence in the context of the response of
an infinitely long system to a localized source. It could perhaps
be of significance in a finite system, however, as is briefly discussed in Section 2.6.

It is important to bear in mind that only the saddle points of
$\omega(k)$ that correspond to a merger of roots from different halves
of the complex k-plane indicate an absolute instability. The following discussion presents a physical interpretation for this re-
striction.

Imagine an infinite system excited by a source that is a spatial
impulse at z = 0. If the source has a complex frequency with the
imaginary part of this frequency larger than the growth rate of any
unstable waves in the system, the waves must all decay away from
the source, as pointed out before (Figure 2.17a). This identifies
which waves appear for $z > 0(k_+)$ and which waves appear in the
response for $z < 0(k_-)$. If we visualize the growth rate of the

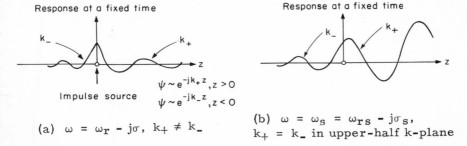

Response at a fixed time

k_-
k_+
z
Impulse source

$\psi \sim e^{-jk_+ z}, z > 0$
$\psi \sim e^{-jk_- z}, z < 0$

(a) $\omega = \omega_r - j\sigma$, $k_+ \neq k_-$

Response at a fixed time

k_- k_+
z

(b) $\omega = \omega_s = \omega_{rs} - j\sigma_s$,
$k_+ = k_-$ in upper-half k-plane

Figure 2.17. Resonance condition in an infinite system.

source as decreasing, then for some complex $\omega = \omega_{rs} - j\sigma_s$ the situation might arise where k_+ and k_- become equal. Clearly, the presence of a spatial impulse type of source must cause a discontinuity in the response or one of its spatial derivatives at $z = 0$. When $k_+ = k_-$, however, we can form a response that does not have any discontinuities and can be smoothly joined across $z = 0$ (Figure 2.17b). This situation is therefore a type of spatial "resonance" of this infinite system at that particular frequency (ω_s) because its presence does not require a source. This is precisely what the analysis in Section 2.3 indicated. If we excite the system with a pulse in time rather than a sinusoid, $f(\omega)$ is an entire function of ω in Equation 2.15, and the response of the system is determined by the singularities of $F(\omega, z)$ in the ω-plane. All of the singularities of $F(\omega, z)$ (except the essential singularities arising from the infinitely short wavelength "resonances" discussed in Section 2.3.4) occur at precisely the frequencies for which we have such a "joining" of k_+- and k_--type waves, since this "joining" is precisely the same as the statement that two roots of the dispersion equation merge through the C contour, as was discussed in Section 2.3.4. In fact, it was shown in that section that the "steady-state response" $F(\omega,z)$ is the same function of z for $z > 0$ and $z < 0$ for $\omega \simeq \omega_s$, in agreement with this physical picture of a "resonance" of the infinite system.

2.5.3 Application of the Criteria in Simple Cases. As was mentioned in the previous discussions, the application of these criteria requires in general that a rather complete conformal mapping of the function $\omega(k)$ be carried out. Unfortunately, this usually requires extensive numerical computations; however, in some cases a few shortcuts can be applied which ease the labor. For instance, if it is known that no complex ω in the lower-half ω-plane are obtained for real k, then there is clearly no possibility of either amplifying waves or absolute instabilities.

Another technique which is sometimes useful is that of analyzing the dispersion equation for complex $\omega = \omega_r + j\omega_i$ with $\omega_i \rightarrow -\infty$; this limit often allows an explicit solution for all possible k's even for higher-order dispersion equations. As a simple illustration of this technique, consider the dispersion equation for longitudinal oscillations in a one-dimensional beam-plasma system. For a cold beam and a cold plasma, this is (see Section 3.1)

$$\Delta(\omega, k) = 1 - \frac{\omega_{po}^2}{\omega^2} - \frac{\omega_{pb}^2}{(\omega - kv_0)^2} = 0 \qquad (2.39)$$

For $\omega < \omega_{po}$, there are complex roots for k. If $|\omega| \rightarrow \infty$, the roots for k are

$$k \simeq \frac{\omega \pm \omega_{pb}}{v_0} \qquad (2.40)$$

and therefore both roots are in the lower-half k-plane when $\omega_i \to -\infty$. This means that both roots of k are waves that enter in the "downstream" side ($z > d$) of any source (as is physically reasonable in this case); consequently, the solution with $k_i > 0$ for real ω represents an amplifying wave. In addition, for the same reason, there is no possibility of an absolute instability.

We note also, in passing, that an essential singularity of $F(\omega, z)$ (for $z > d$) arises at $\omega = \pm\omega_{p0}$, since $\omega^2 \to \omega_{p0}^2$ as $k \to \infty$. This means that the asymptotic response on the "downstream" side of a source can contain undamped plasma oscillations at ω_p as well as at the driving frequency of the source in this zero-temperature model.

In more complicated cases, this technique of analyzing the dispersion equation at infinite frequency will often give some information even if it does not provide a complete answer. For instance, if there are two sets of complex conjugate roots of k for some real ω and it can be determined by letting $\omega_i \to -\infty$ that three enter for $z > 0$, say, and one enters for $z < 0$, it must be true that at least one of the roots is amplifying for $z > 0$, even though it is not clear which one. Further illustrations of some useful techniques for applying the criteria are given in Section 2.7.

2.6 Discussion

2.6.1 Group Velocity of Propagating Waves.
The criterion on amplifying and evanescent waves fixes the "sense of causality" of a wave with a complex wave number k; that is, it determines the direction in which the signal is transferred when the wave is excited at some fixed reference plane. If the criterion is applied to pure propagating waves, some interesting facts about the concept of group velocity become evident.

For propagating waves, k is real for real ω. If we add an infinitesimal negative imaginary part to the real ω, the root of the dispersion equation moves out into the complex k-plane in a direction depending on the sign of $\partial\omega/\partial k$. If the system is stable, that is, if no complex ω with negative imaginary parts for real k are obtained from the dispersion equation, then this root can never cross the real-k axis again as the (negative) imaginary part of ω is increased. In this case, the sign of $\partial\omega/\partial k$ correctly gives the direction of propagation of the wave, as we should expect. This statement is not necessarily true if the system supports unstable waves, however, since the roots can then cross the real-k axis again for some ω with a finite negative imaginary part.

There are many examples of unstable systems in which the sign of $\partial\omega/\partial k$ does not give the correct direction in which the signal is transferred in a propagating wave. As one example, we again consider the case of longitudinal oscillations in a cold beam-plasma system of infinite extent, as described in Section 3.1. In Figure 3.2, we shall see that there is a region where $\partial\omega/\partial k < 0$ for real ω slightly larger than ω_{po}. According to the discussion in Section 2.5, however, both roots of k are waves that enter in the "downstream" side when the system is excited by a localized source. As a matter of interest, the locus of this root in the k-plane as the negative imaginary part of ω is varied is plotted in Figure 2.18 for several values of real ω. We see that the propagating wave with $\partial\omega/\partial k < 0$ does indeed move into the upper-half k-plane for small negative imaginary part of ω; however, it eventually crosses the real-k axis for sufficiently large negative imaginary part of ω.

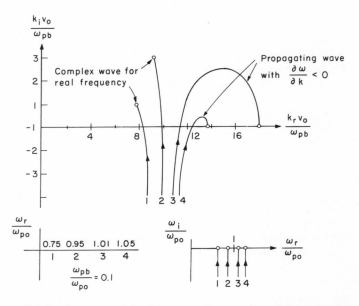

Figure 2.18. Locus of k for one-dimensional beam-plasma system (see Figure 3.2).

This breakdown of the concept of group velocity in a system that supports unstable waves is also reasonable from the more usual picture of the propagation of a pulse on the system. The point is that the spatial Fourier spectrum of the pulse, if the pulse is finite in extent, extends over all real wave numbers, so that the dominant contribution as t → ∞ comes from regions where complex ω for real k are obtained and not necessarily from the peak in the quasi-monochromatic spectrum of k at t = 0. The logical extension of the concept of group velocity to systems that support unstable waves (as regards the propagation of a spatial pulse) was given in Section 2.4.

2.6.2 Comparison with Previous Work. The criteria derived in this chapter differ in one way or another from the criteria that have been previously published (see References 45, 46, 51-53).

All of these works, including the present formulation, have re-
lied upon an investigation of only the dispersion equation to establish the criteria on amplifying waves and absolute instabilities.
For this reason, it is of interest to compare briefly these criteria with the present formulation.

In Sturrock's pioneering work,[46] he recognized the fact that the
dispersion equation should contain the necessary information, and
he established the method of looking at pulses in space or time
which was followed by later authors. (A brief indication of this
general approach in the case of distinguishing between absolute
and convective instabilities was given by Landau and Lifshitz.[45])
Even in the context of his formulation, however, Sturrock did not
carefully consider the effects of the branch points of $\omega(k)$ and $k(\omega)$
when deforming contours in the complex k and ω planes, as was
pointed out in References 51 and 52. These singularities are of
importance in any dispersion equation which is of higher order
than first in ω and k; for this reason there is little correspond-
ence of Sturrock's result with the results of the present formulation.

Fainberg, Kurilko, and Shapiro[51] consider the asymptotic behavior of a disturbance that is initially in the form of a spatial
pulse, in order to distinguish between absolute and convective instabilities. The criteria they obtain state that an absolute instability results whenever there is a saddle point of $\omega_\alpha(k)$ in the lower-
half ω-plane and between the contour of complex $\omega_\alpha(k)$ for real k
and the real-ω axis (Figure 2.19). Here ω_α is one of the branches
(sheets) of the function $\omega(k)$, and it is important in their criteria

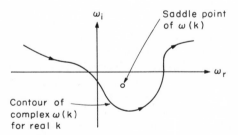

Figure 2.19. Criteria of Fainberg, Kurilko, and Shapiro.[51]

that the saddle point being considered lies on the same sheet of
$\omega(k)$ as the contour. This criterion is similar but not identical
to the one derived in the present work; the requirement that the
saddle point be inside the contour of complex ω for real k and
the real-k axis, and on the same sheet, is almost the same as
the requirement that one of the roots of $\Delta(\omega, k) = 0$ in the k-plane
cross the real-k axis and merge into a double root of k for some

ω as we bring the frequency up from the Laplace contour. If the root crosses the real-k axis <u>twice</u>, however, there is a possibility that the two criteria are <u>not</u> in agreement. (Actually, if it crosses an <u>even</u> number of times, the two criteria might disagree.) This is discussed in more detail in Appendix A, where it is shown that the difference in the two formulations arises because the <u>branch</u> points of $\omega(k)$ were not carefully considered in Reference 51. In this appendix, a rather artificial numerical example is discussed to illustrate these ideas. (Note also that the problem of distinguishing amplifying and evanescent waves is not considered in Reference 51.)

Polovin[52] has considered the question of distinguishing amplifying and evanescent waves from a basic formulation which is very similar to the one proposed by Sturrock.[*] A perturbation is considered that is assumed to be a pulse in <u>time</u> at some fixed point in space. If this perturbation vanishes for $z \rightarrow \pm\infty$, then the wave is defined as amplifying; otherwise the wave is defined as evanescent. The resulting criterion states that if there is a saddle point of $k(\omega)$ inside of the contour of complex k for real ω and the real-k axis, and on the same sheet, the wave is evanescent; otherwise, it is an amplifying wave. This mathematical criterion does not bear any resemblance to the one derived in the present work. It is believed that the difference between these two criteria arises from the difference in the basic <u>physical definition</u> of amplifying and evanescent waves. As a counterexample to the Polovin criterion, consider the dispersion equation derived for forward-wave interaction in a weak-coupling approximation (see Section 2.7, Equation 2.44):

$$\left(k - \frac{\omega}{V_1}\right)\left(k - \frac{\omega}{V_2}\right) = -k_0^2 \tag{2.41}$$

It is easily shown that all saddle points of $k(\omega)$ in this case are on the <u>real</u>-k axis, and therefore Polovin's criterion would state that <u>both</u> complex roots of k for real ω are amplifying waves. This is at variance with the well-known results of both theory and experiments on traveling-wave-tube amplifiers.

It should also be mentioned that Buneman has developed a criterion on amplifying and evanescent waves that involves determining the admittance of a probe inserted into the system.[53] In the analysis, however, he considers simultaneously an infinite system and purely real frequencies, a procedure that is subject to the criticisms given in Section 2.2. He also, clearly, does not consider the problem of distinguishing absolute and convective instabilities.

---

[*]Most of Polovin's work is concerned with interpreting the two-wave weak-coupling diagrams like those discussed in Section 2.7. Only his basic formulation of the problem is of interest in this critique.

2.6.3 Usefulness of Criteria. The criteria that we have derived on amplification and instability were based on a model of
an infinite system driven by a localized source. The reason for
doing this was that we wished to obtain information about the waves
on the uniform system without reference to any particular set of
boundary conditions in z. Having done this, however, it is important to clarify exactly what information has been obtained by
this procedure in regard to the behavior of systems of finite length.
In this respect, it is perhaps worth while first to give some warn-
ings about things that our criteria do not say about system of finite
length.

1. One should not be led into the incorrect conclusion that a
system which supports an absolute instability is always unstable.
Such a system usually must be longer than some critical length
before the system becomes unstable, as for example, in the case
of the backward-wave oscillator.[58] The statement that a system
which supports absolute instabilities is usable only as an oscillator is also not true. A well-known counterexample is that of
the backward-wave amplifier;[2] the absolute instability is suppressed in this case by making the tube short enough, and yet a
net gain can be realized by operating close to the oscillation condition.

2. It is also not true that a system of finite length which supports only convective instabilities (amplifying waves) is necessarily stable, that is, that perturbations cannot also grow in time
at every point in space. All this statement really says is that an
amplifier can become an oscillator if there is a sufficient reflection of an amplifying wave, arising from terminations of the system, as, for example, in the case of mismatches at the output
and input of a traveling-wave-tube amplifier.

It is believed that these criteria provide very useful information, however, as long as one exercises due caution when applying the results to physical situations. To list a few of the positive statements that can be made:

1. The criteria on amplifying and evanescent waves tell one
if it is ever possible to have exponential spatial amplification of
a signal, without considering a number of boundary-value prob-
lems in z, that is, without considering the "coupling of the signal" in and out of the system.

2. If the uniform system supports absolute instabilities, then
temporally growing oscillations can occur without the necessity
of any reflections from terminations or external feedback. In
many physical situations, it is obvious that no mechanism for such
reflections (or reverse propagation) exists, in which case the
"stability" of the finite system can be determined almost directly
from these criteria (except for the "starting length" question al-
ready discussed).

3. The "propagation velocity" defined in Section 2.4 is useful
for estimating whether a convective instability would grow to large
amplitude before reaching the ends of the system. For many e-
folding times to have evolved, the length of the system L should be
much larger than V_0/σ_0, where V_0 and σ_0 are defined in Section 2.4.
It is interesting to note, in this regard, that the lower bound that
was derived on the maximum amplification rate α_M (Inequality 2.36)
shows that the above-mentioned condition requires also that $\alpha_M L \gg 1$
This is another example of how the analysis in Section 2.4 allows
one to connect the concepts of a growing pulse disturbance with the
amplification in space of a sinusoidal signal.

In brief, the criteria derived here supply additional information
about "potentially unstable" systems beyond the simple requirement that complex ω with negative imaginary part for real k be
obtained from the dispersion equation of the infinite (uniform) system. In addition, this development has clearly demonstrated that
the problem of distinguishing amplifying and evanescent waves can
never be separated from the question of stability. This would also
be true whenever systems of finite length are considered.

2.6.4 Amplifying Waves in the Presence of an Absolute Instability. It is of interest to review briefly the example of an absolute
instability presented in Section 2.4 in the light of the previous discussion on systems of finite length. In the case illustrated in Figure 2.16, as was mentioned before, there is an abrupt transition
of the root that has complex k for real ω from amplifying to evanescent for some real ω in the vicinity of (the real part of) the
frequency of the absolute instability. Several physical examples
exhibiting this behavior will be presented in Chapters 4 and 5. In
the context of an infinite (uniform) system, of course, this is of
no significance because there is an absolute instability present.
Difficulties in interpretation do arise, however, in regard to the
behavior of systems of finite length. In a system that is too short
for the absolute instability to be excited, as, for example, a
backward-wave amplifier,[2] it is meaningful to speak of real frequencies. Since the frequency at which this "transition" occurs
is not unique,* the criteria based on the infinite system are not
able to give any definitive information concerning the meaning of
these roots, "amplifying" or "evanescent," in the finite system
in this case. The only general conclusion possible in the context
of the present work is that, to answer this question, the system
of finite length must be considered explicitly in such cases.

---

*Note that the branch line of $F(\omega, z)$ can be oriented at an arbi-
trary angle in the ω-plane, and not just at 90° to the real-ω axis
as shown in Figure 2.11. It can be shown that this will make a
corresponding alteration in the real frequency at which this transition from "amplification" to "evanescence" occurs.

2.7 Examples

In this section, some simple examples of the application of the criteria are treated in order to illustrate the techniques involved. In the first part the quadratic equations obtained from a coupled-mode description of weakly coupled systems are analyzed and discussed, while the second set of examples is drawn from the field of plasma physics.

2.7.1 Weak-Coupling Dispersion Equations.

The operation of most beam-type microwave tubes can be understood by a weak-coupling formalism.[48,49,59] In this section we apply the criteria on amplifying waves and absolute instabilities to the dispersion equations that result from such a weak-coupling formulation. We shall consider only the coupling of two propagating waves in lossless systems; the four possible types of dispersion diagrams that can result are shown in Figure 2.20. (Trivial variations in these

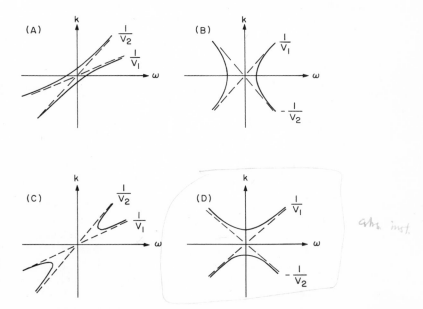

Figure 2.20. Weak-coupling dispersion diagrams.

diagrams are not considered as different types.) The equations describing these diagrams in the vicinity of the intersection of the two uncoupled waves are the following:

$$(A) \qquad \left(k - \frac{\omega}{V_1}\right)\left(k - \frac{\omega}{V_2}\right) = k_0^2 \qquad\qquad (2.42)$$

$$(B) \qquad \left(k - \frac{\omega}{V_1}\right)\left(k + \frac{\omega}{V_2}\right) = -k_0^2 \qquad\qquad (2.43)$$

(C) $\left(k - \dfrac{\omega}{V_1}\right)\left(k - \dfrac{\omega}{V_2}\right) = -k_0^2$ (2.44)

(D) $\left(k - \dfrac{\omega}{V_1}\right)\left(k + \dfrac{\omega}{V_2}\right) = k_0^2$ (2.45)

In these equations, for algebraic simplicity, the point of intersection of the two uncoupled waves has been shifted to the origin
of the $(\omega - k)$-plane. We consider each of these cases separately:

(A). In this case k is real for all real ω, and conversely ω is
real for all real k. Therefore, there are no instabilities, and
the resulting coupled waves are "ordinary" propagating waves.

(B). Now there are complex roots of k for real ω, but ω is real
for all real k. There are no unstable waves, and hence we can
immediately conclude that the complex waves for real ω are both
evanescent waves.

(C). In this case there are complex roots for k for real ω and
complex roots of ω for real k. As $|\omega| \to \infty$, the two roots be-
come   *amp. of evanescent*

$$k \to \frac{\omega}{V_1}$$ (2.46)

and

$$k \to \frac{\omega}{V_2}$$ (2.47)

and therefore both roots are in the lower-half k-plane for $\omega_i \to -\infty$.
It follows that the root with $k_i > 0$ for real ω is an amplifying wave
in the $+z$ direction and the root with $k_i < 0$ is an evanescent wave.
There is also no possibility of an absolute instability because both
waves come from the lower-half k-plane and therefore they cannot
merge through the contour C (the deformed Fourier contour). Moreover, it is easily shown that all double roots of k occur for real ω.

As a matter of interest, the reader can check for himself that
the lower and upper bounds on the maximum amplification rate of
a convective instability derived in Section 2.4 are satisfied for this
simple example.

(D). Now k is real for all real ω; however, there are complex
roots of ω for real k. By solving the quadratic for k, we find
that a double root of k occurs when

$$\omega = \omega_s = -2jk_0\left(\frac{1}{V_1} + \frac{1}{V_2}\right)^{-1}$$ (2.48)

As $|\omega| \to \infty$, the roots become

$$k \to \frac{\omega}{V_1} \qquad (2.49)$$

and

$$k \to - \frac{\omega}{V_2} \qquad (2.50)$$

and therefore for $\omega_i \to -\infty$, one root is in the upper-half k-plane and one root is in the lower-half k-plane. We now have an absolute instability at the frequency given by Equation 2.48, and at the wave number

$$k = k_s = jk_0 \left(\frac{1 - \dfrac{V_2}{V_1}}{1 + \dfrac{V_2}{V_1}} \right) \qquad (2.51)$$

It is interesting to note also that the spatial pattern of the asymptotic response in this case (which is given by $e^{-jk_s z}$ with k_s given by Equation 2.51) is exponentially increasing in the direction of the "faster wave" (see Equation 2.22). That is, when $V_1 > V_2$, this spatial pattern has an exponential envelope increasing in the direction taken by the group velocity of wave 1 when the waves are uncoupled.

As a matter of interest, the loci of the roots in the k-plane for these four cases are sketched in Figure 2.21 for complex ω with $\omega_r = 0$. Note, however, that it was not necessary in these simple cases to perform a detailed mapping of $\omega(k)$ in the complex k-plane, as would be the case in more complicated situations.

All of these results are in agreement with the predictions based on the concepts of small-signal energy and power.[49,59] Diagrams A and B are of the type that result from a weak coupling of two passive waves or from a coupling of two active waves. By "passive wave," we mean a wave that has positive small-signal energy, and the term "active wave" means that the wave carries negative small-signal energy. Diagrams C and D are the type that result from a coupling of an active wave with a passive wave. When the group velocities of the uncoupled waves are in the same direction, as in C, we have amplification as, for instance, in a traveling-wave amplifier.[47] When the group velocities of the uncoupled waves are in opposite directions, as in D, an absolute instability results, as, for example, in a backward-wave oscillator.[58]

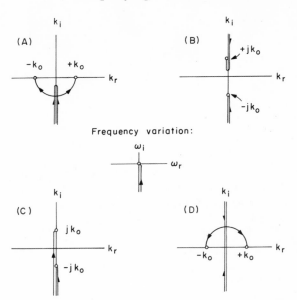

Figure 2.21. Loci of roots in k-plane for weak-coupling examples ($\omega_r = 0$ and $V_1 > V_2$).

2.7.2 Double-Stream Interactions. We now consider a dispersion equation that describes one type of the well-known electrostatic (streaming) instabilities in a plasma. This example is less trivial than the previous ones, and illustrates some additional techniques for applying the criteria without resorting to detailed numerical calculation.

The dispersion equation we consider is

$$1 = \frac{\omega_{ps}^2}{(\omega - kv_0)^2} + \frac{\omega_{ps}^2}{(\omega + kv_0)^2} + \frac{\omega_{pp}^2}{\omega^2} \qquad (2.52)$$

This equation describes the longitudinal waves in a system of charged particles with two equal-density ("cold") streams, each with plasma frequency ω_{ps}, which flow against each other with equal and opposite velocities v_0 and are immersed in a background of stationary cold particles with the plasma frequency ω_{pp} (see Appendix C). This is a generalization of the dispersion equation (2.39) which described a single stream flowing through a plasma.

If we define

$$K = 1 - \frac{\omega_{pp}^2}{\omega^2} \qquad (2.53)$$

then Equation 2.52 can be solved for k^2 as

$$(kv_0)^2 = \omega^2 + \frac{\omega_{ps}^2}{K} \pm \frac{\omega_{ps}}{K} (4\omega^2 K + \omega_{ps}^2)^{\frac{1}{2}} \qquad (2.54)$$

We consider first the special case of a double-stream inter-action in the absence of the background plasma ($\omega_{pp} = 0$ or $K = 1$). In this case, a double root of k, saddle point of $\omega(k)$, does occur in the lower-half ω-plane for

$$\omega = \omega_s = -j \frac{\omega_{ps}}{2} \qquad (2.55)$$

To show that this saddle point corresponds to an absolute in-stability, we sketch in Figure 2.22 the root loci in the k-plane for pure imaginary ω. As shown in the figure, it is convenient in this example to sketch the loci in the k^2-plane first and then to derive the k-plane loci from the k^2 loci. As is evident from the figure, we have an absolute instability in this case at the pure imaginary frequency ω_s.

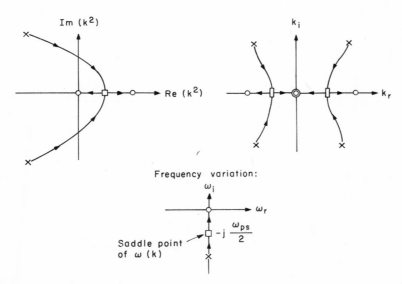

Figure 2.22. Double-stream instability without a stationary plasma ($\omega_{pp} = 0$).

In the case that includes a stationary plasma, we find from Equation 2.54 that a double root of k for an ω in the lower-half plane occurs only when

$$\omega_{pp} < \frac{\omega_{ps}}{2} \qquad (2.56)$$

When Inequality 2.56 is satisfied, the saddle point occurs at the frequency

$$\omega_s = -j \frac{\omega_{ps}}{2}\left(1 - \frac{4\omega_{pp}^2}{\omega_{ps}^2}\right)^{\frac{1}{2}} \tag{2.57}$$

It can be shown, in a similar manner as for the special case $\omega_{pp} = 0$, that when Inequality 2.56 is satisfied the saddle point given by Equation 2.57 does correspond to an absolute instability at ω_s. On the other hand, when Inequality 2.56 is <u>reversed</u>, we know that there cannot be an absolute instability because there is no saddle point in the lower ω-plane. It is readily shown from Equation 2.54 that there are always complex roots of ω for real k, a result indicating the presence of convective instability.

To determine the possible amplification rates in the case $\omega_{pp} > \omega_{ps}/2$, the complex k values for real ω are sketched in Figure 2.23. In order to identify which complex roots correspond to am-

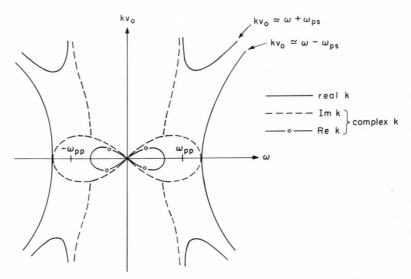

Figure 2.23. Complex k for real ω $(\omega_{pp} > \omega_{ps}/2)$.

plifying waves and which ones correspond to evanescent waves (realizing that the analysis in Section 2.4 has <u>proved</u> that amplifying waves must exist in this case), we might perform numerical computations on a particular case. It is perhaps more instructive, however, to use the somewhat intricate but more generally useful argument now to be outlined.

We shall show that the two roots of k that have $k_i \to \pm\infty$ as $\omega \to \omega_{pp}$ are amplifying waves; that is, there is an infinite amplification rate predicted by this model in <u>both</u> the +z and -z directions when $\omega_{pp} > \omega_{ps}/2$. To demonstrate this, it is convenient to trace the k-plane loci for a rather devious route for ω (Route a in Figure 2.24b) rather than the more standard route obtained by

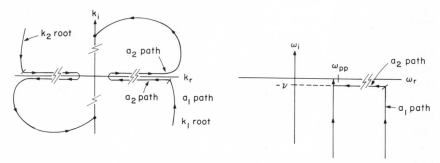

(a) Loci of two of the roots of k.   (b) Frequency variations.

Figure 2.24. Loci of k roots ($\omega_{pp} > \omega_{ps}/2$). The roots of k follow the trajectories indicated in (a) as the frequency varies along Route a in (b).

holding the real part of ω fixed on the desired real frequency and bringing ω_i from $-\infty$ to zero (Route b). The validity of this approach is discussed in the following paragraph. The advantage of traveling along Route a is that we can bring ω up to a point just below the real axis ($\omega = \omega_r - j\nu$), and at the same time keep $|\omega|$ very large in the process, so that we <u>know</u> which root of the fourth-order equation (2.54) we are dealing with (Path a_1) in Figure 2.24a. In the figure, the two roots indicated in the k-plane are those that become

$$k_1 v_0 \simeq \omega + \omega_{ps} \tag{2.58}$$

$$k_2 v_0 \simeq -\omega + \omega_{ps} \tag{2.59}$$

as $|\omega| \to \infty$. We then bring ω_r down from large positive values to the desired real frequency (slightly below ω_{pp} in the present case), using the real ω diagram to follow the root of interest (Figure 2.23). In this process, we retain the infinitesimal negative imaginary part of ω in order to avoid going directly through any "indeterminate" points like ω_{pp}. For the reasons discussed in the following paragraph, the use of this small imaginary part of ω tells us how to join correctly the k roots on the opposite sides of such singular points. When this process is carried out, we find that the two roots of interest have performed the trajectories shown in Figure 2.24, and this verifies the result stated earlier.

It still remains to determine the conditions for which this device "works"; namely, under what conditions is it true that the "same answer" is obtained by going along Path a as would be obtained by going along Path b? A little thought will convince us that the only way we could reach at a different point in the k-plane for the same real value of ω when traveling different paths (which both start with the same root of k and at the same point, namely, $\omega_i = -\infty$) is if a branch point of $k(\omega)$ is enclosed within these two paths in the ω-plane. If we do not enclose such a branch point, then when we start on a particular sheet of $k(\omega)$ (particular root of k), we shall be on the same sheet when we arrive at the end point, even if different paths are chosen. All this really says is that if we were to go through a point that is a double root of k — a branch point of $k(\omega)$ — we would "lose track" of which root we are following. If there are no such double roots of k in the lower-half ω-plane, as is the case here, then the procedure is valid. There are branch points of $k(\omega)$ on the real-ω axis in this example, however, such as ω_{pp}, and this is the reason for retaining a small negative imaginary part of ω when traveling along Path a_2. We emphasize again that this "trick" is useful in sorting out which waves are amplifying only when it is known that no double roots of k occur in the lower-half ω-plane.

The physical results we have obtained are interesting and worth a brief discussion. We showed in Section 2.5 that a single stream in a plasma results in an infinite amplification rate in the direction of the stream and no absolute instability. Two colliding streams in the absence of a plasma, on the other hand, were shown in this section to result in an absolute instability, which is quite reasonable physically since the system has "built-in feedback." When two streams collide in the presence of a plasma, however, the behavior obtained depends on the relative density of the streams to the plasma (Inequality 2.56). For a sufficiently tenuous plasma, the absolute instability between the streams is obtained, whereas a denser plasma results in infinite amplification in both stream directions and no absolute instability.

Chapter 3

BEAM-PLASMA INTERACTIONS IN A
ONE-DIMENSIONAL SYSTEM

In this chapter we shall consider the various types of interactions that can arise in a beam-plasma system of infinite extent.
As mentioned in the Introduction, the majority of the theoretical
works on beam-plasma interactions have used the one-dimensional
model because of its simplicity. The strict applicability of this
model to a given physical situation requires that the transverse
dimensions of the system be much larger than the wavelength $2\pi/k$,
which is almost never the case for phenomena of laboratory scale.
However, the simplicity of this model does allow for a fairly complete description of the finite-temperature plasma, and much of
the basic physics of the interactions is brought out without the complications introduced by the finite transverse boundaries.

In the past, most of the effort has been concentrated on the analysis of the relatively high frequency interactions with the plasma
electrons, where the presence of the ions can be ignored.[12,13] One
of the primary purposes of the investigation in this chapter is to
determine the possibility of a strong interaction with the plasma
ions.* To determine the conditions for which this might occur, the
interactions at both high and low frequencies will be analyzed over
a wide range of electron temperature.

In this one-dimensional model it is assumed that the beam-plasma
system is homogeneous and of infinite extent in the plane transverse
to the electron-beam flow. The beam electrons move with the unperturbed velocity v_0 that is constant and parallel to an externally
applied magnetic field B_0 (Figure 3.1). Since the primary interest

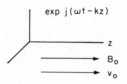

Figure 3.1. One-dimensional system.

---

*The possibility that an electron beam might interact with the
ions of a plasma when the plasma electrons are relatively warm
was first suggested by Professor L. D. Smullin of M.I.T.

is in the case of an externally generated electron beam that is injected into the plasma, it will be assumed throughout that the beam is cold, that is, that all beam electrons have the same unperturbed velocity. Only the waves propagating along the magnetic field will be considered; for this case the space-time dependence of all variables in the linearized approximation can be taken to be of the form exp $j(\omega t - kz)$.

Furthermore, it is assumed throughout that the plasma particles are adequately described by the collisionless Boltzmann-Vlasov equation. In this approach, the essential assumption is that the particles interact only through the long-range, macroscopic electromagnetic fields. The dispersion equation that results from such a treatment is well known;[26] for completeness it is derived in Appendix C from a "multibeam" or "particle-orbit" analysis.

In Appendix C it is shown that the waves along the magnetic field split into two uncoupled types: (1) longitudinal waves, with only a longitudinal component of the electric field E_z and zero magnetic field $\overline{H} = 0$; and (2) transverse waves, which have $\overline{E}$ and $\overline{H}$ perpendicular to $\overline{B}_0$. The transverse waves themselves split into uncoupled right- and left-circularly polarized waves. It is clear physically that the longitudinal waves are independent of the presence of the steady magnetic field $\overline{B}_0$, and therefore this dispersion equation is just the well-known result for the case of $\overline{B}_0 = 0$.

A. LONGITUDINAL INTERACTIONS

3.1 Cold Plasma

The dispersion equation for the longitudinal interaction in a cold plasma is[12,13]

$$1 - \frac{\omega_{po}^2}{\omega^2} - \frac{\omega_{pb}^2}{(\omega - kv_0)^2} = 0 \tag{3.1}$$

where $\omega_{po}^2 = \omega_{pi}^2 + \omega_{pe}^2$ is the total plasma frequency. It was shown in Section 2.5 that the preceding dispersion equation (3.1) predicts that the root with $k_i > 0$ is an amplifying wave and that there are no absolute instabilities in this case. The solution of complex k for real ω is shown in Figure 3.2a. This cold-plasma model predicts an infinite amplification rate at $\omega = \omega_{po}$, as is well known.[5]

The beam waves in the absence of the plasma are given by

$$k = \frac{\omega \pm \omega_{pb}}{v_0} \tag{3.2}$$

and these are sketched in Figure 3.2b. These beam waves are obviously strongly perturbed by the presence of the plasma. In term

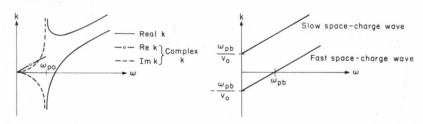

(a) Beam-plasma dispersion.   (b) Beam waves in the ab-
                                    sence of the plasma.

Figure 3.2. Longitudinal interaction with a cold plasma.

of a "coupling" of the beam waves to the plasma waves, we would
describe this longitudinal interaction in a cold plasma as being a
case of "very strong coupling."

A number of authors have considered the effect of a low electron temperature on this interaction.[14,15] As would be expected,
the amplification rate is no longer infinite for finite temperatures;
but for temperatures much less than the beam voltage, it is still
extremely large.

In the limit of very low electron temperatures, this longitudinal
interaction clearly involves only the plasma electrons. For instance, it is easily shown that in a cold plasma the time-averaged
kinetic energy of the plasma electrons is always M/m times the
average energy of the ions for any frequency of excitation. The
physical reason for this is that the plasma electrons tend to "short
out" the fields at low frequencies because of their smaller mass
and thereby prevent an independent ion oscillation from occurring.
If the electron temperature were quite high, however, one might
expect that the situation would be different. For relatively high
electron temperatures, a large number of the electrons are in
relative motion with respect to the laboratory frame, and these
electrons are acted on by a force at the Doppler-shifted frequency
$\omega - kv_e$, where v_e is the unperturbed velocity of a particular electron. For large wave numbers, the magnitude of the frequency
can be appreciably larger than ω. This will lead to less "shorting
out" of a low-frequency electric field because the small-signal
velocity of the electron is inversely proportional to this Doppler-
shifted frequency. One of the principal objectives of the analysis
in the following sections is to determine more precisely the conditions on electron temperature that must be met in order for a
strong interaction with the ions to occur.

3.2 Dispersion Equation for a Warm Plasma

The dispersion equation for longitudinal waves is derived in Appendix C. It can be written in the form

$$1 = \sum_n \omega_{pn}^2 \int_{-\infty}^{+\infty} \frac{f_{0n}(v_z) \, dv_z}{(\omega - kv_z)^2} \tag{3.3}$$

or, performing an integration by parts, as

$$1 = -\sum_n \frac{\omega_{pn}^2}{k} \int_{-\infty}^{+\infty} \frac{\dfrac{\partial f_{0n}(v_z)}{\partial v_z} \, dv_z}{\omega - kv_z} \tag{3.4}$$

where the sum is over all species of charged particles (beam electrons, plasma electrons, and ions), ω_{pn} is the plasma frequency of that species, and $f_{0n}(v_z)$ is the distribution function, which has been integrated over the transverse velocities.

The right-hand side of the dispersion equation (3.3 or 3.4) actually defines two functions of ω and k, depending on whether Im (ω/k) assumed positive or negative when the velocity integration is carried out. Landau has considered the meaning of this for the case of a real value of k impressed on the system.[26,60] In Appendix B, the necessary generalization of Landau's argument to the case of complex k is outlined. In the following, we shall give only the function corresponding to the case where the velocity integration is carried out by assuming that Im $(\omega/k) < 0$, since it turns out that this usually gives the roots of interest for unstable waves in a beam-plasma system (this corresponds to the dispersion equation $\Delta_1(\omega, k) = 0$ defined in Appendix B).

The nonrelativistic Maxwellian distribution has the form

$$f_0(v_z) = \frac{1}{\sqrt{2\pi} \, V_T} e^{-v_z^2/2V_T^2} \tag{3.5}$$

Here, V_T is the average longitudinal thermal velocity; that is,

$$V_T^2 = \int_{-\infty}^{+\infty} v_z^2 f_0(v_z) \, dv_z \tag{3.6}$$

It is related to the temperature by

$$V_T^2 = \frac{eT}{m} \tag{3.7}$$

where T is the temperature in electron volts (for an isotropic Maxwellian distribution) and m is the mass of that species. (The subscript n is suppressed in all of the following.)

Because of the mathematical complexity introduced by the Maxwellian distribution, it is often helpful to work first with various simple distribution functions, to obtain some qualitative feeling for the problem. Two simple forms that will be used in the following sections are the resonance distribution

$$f_0(v_z) = \frac{V_{Tr}}{\pi} \left(\frac{1}{v_z^2 + V_{Tr}^2} \right) \tag{3.8}$$

and the square distribution

$$\left.\begin{aligned} f_0(v_z) &= \frac{1}{2V_{Ts}} \quad \text{for } |v_z| < V_{Ts} \\[2em] f_0(v_z) &= 0 \qquad\quad \text{for } |v_z| > V_{Ts} \end{aligned}\right\} \tag{3.9}$$

The integral in Equation 3.3 for the resonance curve gives

$$I = \omega_p^2 \int \frac{f_0(v_z)\, dv_z}{(\omega - kv_z)^2} = \frac{\omega_p^2}{(\omega - jkV_{Tr})^2} \tag{3.10}$$

for the $\text{Im}\,(\omega/k) < 0$ dispersion equation. Equation 3.10 is obtained most simply by closing the contour in the upper half of the complex v_z-plane. This reduces the integral to the evaluation of the residue at the simple pole at $v_z = jV_{Tr}$, since the double pole at $v_z = \omega/k$ is assumed to be below the contour.

If we consider only one species (say, electrons), then the dispersion equation for longitudinal oscillations for the case of a resonance distribution is, from Equation 3.10,

$$\omega = jkV_{Tr} \pm \omega_p \tag{3.11}$$

(The use of this dispersion equation for real positive k and $\text{Im}\,\omega > 0$ is justified by Landau's argument of analytic continuation.) The waves are therefore damped in time when a real wave number is considered; this is the well-known phenomenon of Landau damping.[60]

The square distribution does not exhibit Landau damping. For this case, we obtain

$$I = \omega_p^2 \int \frac{f_0(v_z)\, dv_z}{(\omega - kv_z)^2} = \frac{\omega_p^2}{\omega^2 - k^2 V_{Ts}^2} \tag{3.12}$$

from a direct integration. It is interesting to note that the dispersion equation resulting from a square distribution is of exactly the same form as that obtained from a transport equation model, where the temperature effects are included by means of an isotropic pressure term in the force equation.[61]

Various approximate forms of the integral I for a Maxwellian distribution will be used in the following analysis. The integration in Equation 3.4 can be written in the form[26]

$$I = - \frac{\omega_p^2}{k} P \int_{-\infty}^{+\infty} \frac{\frac{\partial f_0}{\partial v_z}}{\omega - kv_z} dv_z - j\pi \left(\frac{\omega_p^2}{k^2}\right) \left(\frac{\partial f_0}{\partial v_z}\right)_{v_z = \frac{\omega}{k}}$$

$$(3.13)$$

for the Im $(\omega/k) < 0$ dispersion equation. In the first term the integral is carried out through the pole at $v_z = \omega/k$ and P signifies the principal part of the integration. This form is particularly useful for real ω and real k because it then expresses the integral explicitly in terms of its real and imaginary parts. In the "low-temperature limit," when $\omega/k \gg V_T$, we can approximate the integral as[26]

$$I \simeq \frac{\omega_p^2}{\omega^2} \left(1 + \frac{3k^2 V_T^2}{\omega^2}\right) + j\sqrt{\frac{\pi}{2}} \left[\frac{\omega_p^2 \omega}{(kV_T)^3}\right] e^{-\omega^2/2k^2 V_T^2}$$

$$(3.14)$$

The Landau damping term Im I is exponentially small for a Maxwellian distribution when the phase velocity is much greater than the thermal velocity. The square distribution correctly approximates the form of the Re I, but it does not contain the Landau damping term. The resonance distribution also reproduces the form of the Re I, but the Im I is much too large in this limit. This is to be expected, since the resonance curve falls off very slowly for large v_z as compared with a Maxwellian distribution.

In the "high-temperature limit," when $\omega/k \ll V_T$, Equation 3.13 becomes

$$I \simeq \frac{\omega_p^2}{k^2} \int_{-\infty}^{+\infty} \frac{1}{v_z} \left(\frac{\partial f_0}{\partial v_z}\right) dv_z + j\sqrt{\frac{\pi}{2}} \left[\frac{\omega_p^2 \omega}{(kV_T)^3}\right] \simeq - \frac{\omega_p^2}{k^2 V_T^2} \left[1 - j\sqrt{\frac{\pi}{2}} \left(\frac{\omega}{kV_T}\right)\right]$$

$$(3.15)$$

where we have used the relation

$$\frac{\partial f_0}{\partial v_z} = - \frac{v_z}{V_T^2} f_0(v_z) \tag{3.16}$$

for a Maxwellian distribution.

In this limit the Landau damping term Im I is again small compared with the Re I; however, the dependence is linear with ω/kV_T and not exponential as in the "low-temperature" regime. Once again, the square distribution has correctly reproduced the form of the Re I although it does not contain the damping term. The resonance distribution is quite good in this limit, as it gives

$$I \simeq - \frac{\omega_p^2}{k^2 V_{Tr}^2} \left[1 - j2\left(\frac{\omega}{kV_{Tr}}\right) \right] \tag{3.17}$$

for the same approximation. The only difference is a numerical factor of $\sqrt{\pi/8}$ in the Im I compared with the Re I.

In summary, the evaluation of the integral in the dispersion equation of longitudinal waves has been considered for three types of distributions: Maxwellian distribution, square distribution, and a resonance distribution. It was shown that the square distribution neglects the Landau damping but agrees in form with the Maxwellian for ω/k much larger or much smaller than V_T (where we should equate V_{Ts} with $\sqrt{3}\, V_T$ in the former case, and with V_T in the latter case). The resonance curve gives far too much Landau damping for the "low-temperature region" when $\omega/k \gg V_T$; however, it is an excellent approximation to a Maxwellian distribution when ω/k is of the order of, or less than, V_T when we equate V_{Tr} with V_T. Because of the mathematical simplicity of the square- and resonance-type distributions, they will be used extensively to obtain some feeling for the results to be expected. In the remainder of the chapters, for notational simplicity, we shall use the symbol $V_{T(e,i)}$ to stand for V_{Tr} and V_{Ts}. It is hoped that no confusion will arise from this convention.

3.3 Weak-Beam Theory

The general solution for ω and k from the dispersion equation of a beam-plasma system with warm plasma electrons is quite difficult. In this section the case of a beam with vanishingly small density will be analyzed. For such a weak beam the plasma waves are not significantly perturbed unless their phase velocity is very close to synchronism with the beam, that is, unless $\omega/k \simeq v_0$. The beam waves, however, are strongly perturbed by the plasma, since the dielectric constant of the plasma medium is quite different from its free-space value. Sections 3.4 and 3.5 treat cases of "stronger coupling"; however, the simple ideas developed in this section will still prove useful in the interpretation of those results.

The dispersion equation for the beam-plasma system can be written in the form (Equation 3.3):

$$\frac{\omega_{pb}^2}{(\omega - kv_0)^2} = K_{\parallel}(\omega, k)$$

(3.18)

where $K_{\parallel}(\omega, k)$ is the longitudinal dielectric constant of the plasma in the absence of the beam.[26] For cold ions and warm electrons, it takes the form

$$K_{\parallel}(\omega, k) = 1 - \frac{\omega_{pi}^2}{\omega^2} - \omega_{pe}^2 \int_{-\infty}^{+\infty} \frac{f_{0e}(v_z)\, dv_z}{(\omega - kv_z)^2}$$

(3.19)

For a very weak beam ($\omega_{pb} \to 0$), some of the solutions of Equation 3.18 are just $K_{\parallel}(\omega, k) \simeq 0$ unless $\omega \simeq kv_0$, that is, the stable plasma waves in the absence of the beam. The solution for the beam waves, which are the remaining roots of Equation 3.18, can be obtained by writing

$$\omega - kv_0 = \delta_\omega(k)$$

(3.20)

and by assuming $\delta_\omega \ll kv_0$. In this formalism, we consider $\delta_\omega(k)$ as the "correction" in ω at some real k because of the (infinitesimal) beam density. We can solve for $\delta_\omega(k)$ from

$$\frac{\omega_{pb}^2}{\delta_\omega^2} = K_{\parallel}(\omega = kv_0) + \delta_\omega \left(\frac{\partial K_{\parallel}}{\partial \omega}\right)_{\omega = kv_0}$$

(3.21)

If the beam is not in synchronism with the plasma waves, that is, if $K_{\parallel}(\omega = kv_0) \neq 0$, the second term in Equation 3.21 can be neglected. From the discussion in Section 3.2, we realize that in general

$$K_{\parallel}(\omega = kv_0) = K_R - jK_I$$

(3.22)

where K_R and K_I are given directly in the form of Equation 3.13, since ω and k are considered real (and positive) in Equation 3.22 The imaginary part of $K_{\parallel}$ arises from the Landau damping. An instability occurs whenever a complex δ_ω is obtained from Equation 3.21 with Im $\delta_\omega < 0$. We can distinguish two basic mechanisms that can cause such an instability:

1. Reactive-medium instability. If v_0 is such that we can neglect Landau damping, that is, such that $K_I \ll K_R$ for $\omega = kv_0$, then

a complex δ_ω is obtained whenever

$$K_R < 0 \tag{3.23}$$

Physically, this means that the plasma medium, which is essentially lossless in this case, has a negative dielectric constant or that the plasma is "inductive" rather than "capacitive" in its relation between the total (convection plus displacement) current and the electric field. This interaction is the reactive-medium amplification, which has been known for some time in the theory of microwave beam tubes. Its analogue in that field is the situation where closely spaced inductively detuned cavities are placed along an electron beam, as in the Easitron amplifier first proposed by Walker.[62] This analogy was first pointed out by Smullin and Chorney in connection with the interaction of an electron beam with an ion plasma.[1,2]

It is possible to argue on an intuitive basis that the condition $K_R < 0$ should give rise to an instability: When a bunched electron beam passes through a medium with a negative dielectric constant, the electrons in a bunch attract rather than repel, and hence the bunching should be further enhanced. From Equation 3.21, it is clear that the maximum growth rate of this instability occurs near $K_{||} \simeq 0$; this is also understandable because the forces between the electrons in a bunch are inversely proportional to the dielectric constant of the medium ($K_{||}$).

2. Resistive-medium instability. When $K_R > 0$ and K_I is not vanishingly small, the solution of Equation 3.21 again gives one root of δ_ω with Im $\delta_\omega < 0$. The mechanism in this case is that of the resistive-medium type of instability. The slow space-charge wave on the beam is a wave that carries negative small-signal energy;[19,20] when the beam passes through a lossy medium, energy is extracted from this wave, causing it to grow in amplitude.[20] The novel feature in the present case is that the loss in the medium comes about because of the Landau damping.

Of course, it is possible to get a combination of these effects if $K_R < 0$ and K_I is of a reasonable magnitude, so that the plasma medium appears "resistive-inductive" rather than "resistive-capacitive" or pure "inductive." In general, however, since the reactive-medium amplification in a relatively lossless plasma is the strongest type of instability, as we saw in Section 3.1, any loss in the case of $K_R < 0$ is usually only a degrading factor on the instability.

In Appendix D it is shown that all instabilities obtained in this section are convective, and that the solution of the beam waves with $k_i > 0$ for real ω is an amplifying wave, as would be expected, since this corresponds to growth in the direction of the beam flow. To find the growth rate in space, we introduce the "correction" to k (of the beam waves) at some real ω, $\delta_k = k - \omega/v_0$. We then have

$$\frac{\beta_{pb}^2}{\delta_k^2} \simeq K_{\parallel}(\omega = kv_0) + \delta_k\left(\frac{\partial K_{\parallel}}{\partial k}\right)_{\omega=kv_0} \tag{3.24}$$

To determine the range of beam velocities v_0 for which reactive-medium amplification occurs, the regions where $K_{\parallel}(\omega, k) < 0$ are plotted in Figure 3.3. In this plot, a square distribution was used

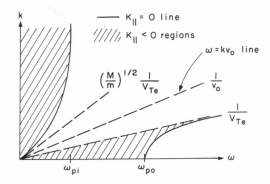

Figure 3.3. Regions where $K_{\parallel} < 0$.

for $f_{0e}(v_z)$ (see Equations 3.12 and 3.19); this neglects the Landau damping and is therefore not valid in the vicinity of $v_0 \simeq V_{Te}$. In the figure, if the straight line defined by $\omega = kv_0$ falls in a region where $K_{\parallel}(\omega, k)$ is negative, an amplifying wave is obtained. It can be shown that the peak amplification occurs at the frequency for which $K_{\parallel}(\omega = kv_0) = 0$.

From the plot we see that for $v_0 > V_{Te}$ there is reactive-medium amplification with a maximum growth rate at a frequency above ω_{po} This is the well-known interaction with the plasma electrons that was considered in Section 3.1 for the case $V_{Te} = 0$; it has been rather thoroughly analyzed and will not be considered further in this investigation. For $(m/M)^{\frac{1}{2}}V_{Te} < v_0 < V_{Te}$, the plasma medium is capacitive ($K_{\parallel} > 0$). There is no reactive-medium instability for this range of beam velocity; however, the inclusion of the Landau damping can result in a resistive-medium type of interaction, as will be discussed later in this section. For $v_0 < (m/M)^{\frac{1}{2}}V_{Te}$, we again have reactive-medium amplification that now has the maximum growth rate at a frequency less than ω_{pi}. We can solve for this frequency from the relation $K_{\parallel}(\omega = kv_0) = 0$. The result is[*]

*The analysis of this interaction using a Maxwellian distribution yields exactly the same results as obtained with a square distribution, since (as was shown in Section 3.2) the dispersion equations are the same for $\omega/k \ll V_{Te}$.

$$\omega^2 = \omega_{pi}^2 \left[1 - \left(\frac{M}{m}\right)\left(\frac{v_0^2}{V_{Te}^2}\right) \right] \tag{3.25}$$

The maximum growth rate follows from Equation 3.24 as

$$(k_i)_{max} = \sqrt{3}\ 2^{-\frac{4}{3}} \left(\frac{\omega_{pi}}{v_0}\right)\left(\frac{n_b T_e}{2 n_p V_0}\right)^{\frac{1}{3}} \left[1 - \left(\frac{M}{m}\right)\left(\frac{2V_0}{T_e}\right) \right]^{\frac{1}{3}} \tag{3.26}$$

From Equation 3.26, we see that a necessary condition for the weak-beam assumption ($|\delta_k| \ll \omega/v_0$) in this range of velocity v_0 is that

$$\left(\frac{n_b}{n_p}\right)\left(\frac{T_e}{2V_0}\right) \ll 1 \tag{3.27}$$

or that $n_b/n_p \ll m/M$ because we are considering the case $2V_0/T_e \lesssim m/M$.

This weak-beam model indicates that a very low velocity electron beam must be used to obtain reactive-medium amplification with the ions of the plasma. In the following section we shall show how the requirement of $2V_0 < (m/M)T_e$ is modified when denser beams are considered.

The discussion so far has neglected the imaginary part of $K_{\parallel}$ that arises from the Landau damping mechanism. We now use a resonance distribution for $f_{0e}(v_z)$ (Equation 3.10) and neglect the second term on the right-hand side of Equation 3.24. The second assumption is allowed within the weak-beam approximation as long as the Landau damping is reasonably large, so that $K_{\parallel}(\omega = kv_0)$ never becomes too small. We then obtain

$$\delta_k^2 = \beta_{pb}^2 \left[\frac{\frac{\omega^2}{\omega_{po}^2}}{\frac{\omega^2}{\omega_{po}^2} - \frac{m}{M} - \left(1 - j\frac{V_{Te}}{v_0}\right)^{-2}} \right] \tag{3.28}$$

It is clear from Equation 3.28 that if $V_{Te} \lesssim v_0$, the maximum Im δ_k will occur at a frequency on the order of ω_{po}; that is, the interaction will again be strongest at high frequencies characteristic of the plasma electrons. For $(m/M)^{\frac{1}{2}} < v_0/V_{Te} \ll 1$, Equation 3.28 becomes approximately

$$\delta_k^2 \simeq \beta_{pb}^2 \left(\frac{\dfrac{\omega^2}{\omega_{po}^2}}{\dfrac{\omega^2}{\omega_{po}^2} + \dfrac{v_0^2}{V_{Te}^2}} \right) \left\{ 1 + 2j \left[\frac{\left(\dfrac{v_0}{V_{Te}}\right)^3}{\dfrac{\omega^2}{\omega_{po}^2} + \dfrac{v_0^2}{V_{Te}^2}} \right] \right\} \tag{3.29}$$

and therefore

$$\operatorname{Im} \delta_k \simeq \beta_{pb} \left(\frac{v_0}{V_{Te}}\right)^3 \left[\frac{\dfrac{\omega}{\omega_{po}}}{\left(\dfrac{\omega^2}{\omega_{po}^2} + \dfrac{v_0^2}{V_{Te}^2}\right)^{3/2}} \right] \tag{3.30}$$

The maximum Im δ_k now occurs at a frequency given by

$$\omega \simeq \frac{1}{\sqrt{2}} \left(\frac{v_0}{V_{Te}}\right) \omega_{po} \tag{3.31}$$

that is, the maximum amplification now occurs below the electron plasma frequency. This maximum amplification is given by

$$\operatorname{Im} \delta_k = \frac{2}{3\sqrt{3}} \left(\frac{\omega_{pb}}{V_{Te}}\right) \tag{3.32}$$

which is much less than β_{pb} for the conditions assumed here.

The amplification rate as a function of frequency is plotted in Figure 3.4 for the case of a Maxwellian distribution. (This amplification rate differs from that obtained from the resonance distribution by only a factor of $\sqrt{\pi/8}$.) In computing the amplification, the approximate expression of Equation 3.15 for the integral of the Maxwellian distribution was used, which assumes that $\omega/k \sim v_0 \ll V_{Te}$. We see that as we go toward smaller values of v_0/V_{Te}, the peak of the amplification rate occurs at lower frequencies, as predicted by Equation 3.31; however, the value of this maximum decreases correspondingly because the Landau damping is very small for these phase velocities.

Summary. In this section the longitudinal interaction of a very weak beam with a plasma was studied. It was found that if $2V_0 \gg$ the strongest interaction is with the electrons at a frequency on the order of ω_{pe}; the mechanism is the well-known reactive-medium type of amplification. If $2V_0$ is on the order of, or slightly less tha T_e, the Landau damping causes a resistive-medium type of amplifi cation with the maximum growth rate at a frequency on the order o $(v_0/V_{Te})\omega_{po}$; however, the value of this growth rate is very small

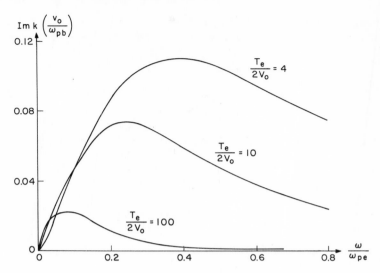

Figure 3.4. Resistive-medium amplification.

for $2V_0 \ll T_e$. Finally, if $2V_0 < (m/M)T_e$, a reactive-medium
type of amplification with the ions at a frequency below ω_{pi} can
result.

All of these results are strictly valid only in the limit of a vanishingly small beam density. In fact, from Equation 3.27, it follows that the low-frequency interaction below ω_{pi} is correctly described by the weak-beam theory only when $n_b/n_p \ll 2V_0/T_e < m/M$.
In the next two sections, the generalization of these results to arbitrary beam densities will be carried out. It will be shown in Section 3.5 that a very strong interaction with the ions can result when
the beam density exceeds a certain critical value.

3.4 General Criteria for a Reactive-Medium Instability with Ions

In the previous section it was shown that a reactive-medium instability with the ions can be obtained for a very weak beam when
$2V_0 < (m/M)T_e$. In this section the condition for such a reactive-
medium instability will be formulated for arbitrary beam densities.
The Landau damping will be neglected for simplicity (and to concentrate on the reactive-medium effects); this allows the method
outlined in Appendix E to be used because the beam-plasma system
is then lossless in this approximation. Basically, this method consists of determining the conditions for which a point of "infinite
group velocity" $(\partial\omega/\partial k = \infty)$ exists for some real ω and real k; if
such a point is obtained, then an instability (complex ω with $\omega_i < 0$
for real k) is present.

3.4.1 Cold Ions. We consider first the case of an electron beam
in a plasma with cold ions and hot electrons (with a Maxwellian distribution). We shall assume that $V_0 \ll T_e$, so that $\omega/k \ll V_{Te}$ in

the range of interest. Using Equation 3.15 to evaluate the term arising from the electrons, we can write the dispersion equation as

$$\Delta(\omega, k) = 1 - \frac{\omega_{pi}^2}{\omega^2} + \frac{\omega_{pe}^2}{k^2 V_{Te}^2} - \frac{\omega_{pb}^2}{(\omega - kv_0)^2} = 0 \qquad (3.33)$$

As shown in Appendix E, the instability criterion requires that

$$\frac{\partial \Delta}{\partial \omega} = 0 \qquad (3.34)$$

for some real ω and k that also satisfy Equation 3.33. From Equation 3.34, we obtain

$$kv_0 - \omega = \left(\frac{\omega_{pb}}{\omega_{pi}}\right)^{\frac{2}{3}} \omega \qquad (3.35)$$

Substituting Equation 3.35 into Equation 3.33, we can solve for the frequency as

$$\omega^2 = \frac{\omega_{pi}^2 \left[1 + \left(\frac{\omega_{pb}}{\omega_{pi}}\right)^{\frac{2}{3}}\right]^3 - \left(\omega_{pe}^2\right)\left(\frac{v_0^2}{V_{Te}^2}\right)}{\left[1 + \left(\frac{\omega_{pb}}{\omega_{pi}}\right)^{\frac{2}{3}}\right]^3} \qquad (3.36)$$

For this frequency to be real, we must have[17]

$$\left(\frac{n_b}{n_p}\right)\left(\frac{T_e}{2V_0}\right)\left[1 + \left(\frac{m}{M}\frac{n_p}{n_b}\right)^{\frac{1}{3}}\right]^3 > 1 \qquad (3.37)$$

Inequality 3.37 represents the desired condition for the appearance of an instability with the ions. In Section 3.5, it will be shown that this instability is very strong when $(n_b/n_p)(T_e/2V_0)$ is itself greater than unity. Note that in the limit of $n_b \to 0$, Inequality 3.37 becomes

$$\left(\frac{T_e}{2V_0}\right)\left(\frac{m}{M}\right) > 1 \qquad (3.38)$$

as obtained in Section 3.3. For the case of protons, with n_p/n_b = 100, Inequality 3.37 requires that

$$V_0 < 0.013T_e \qquad (3.39)$$

a condition that is considerably less restrictive than Inequality 3.38 would give for this case ($V_0 < 2.8 \times 10^{-4}T_e$).

The instability condition (3.37) was derived under the assumption of $T_e \gg V_0$; however, we see that this assumption is consistent in all cases where $n_b \ll n_p$.

3.4.2 Warm Ions. In the previous analysis it was shown that a reactive-medium instability with cold ions can be obtained if the electrons are hot enough and/or the beam density is large enough. In this section the case of a finite ion temperature will be considered. It will be shown that this instability persists as long as $V_{Ti} < v_0$, where V_{Ti} is the average thermal velocity of the ions, when $(n_b/n_p)(T_e/2V_0) > 1$.

A square distribution is used for the ions in order to rule out any resistive-medium effects and to concentrate on the reactive-medium instability. The dispersion equation is then

$$\Delta(\omega, k) = 1 - \frac{\omega_{pi}^2}{\omega^2 - k^2 V_{Ti}^2} + \frac{\omega_{pe}^2}{k^2 V_{Te}^2} - \frac{\omega_{pb}^2}{(\omega - kv_0)^2} = 0$$

$$(3.40)$$

if $v_0 \ll V_{Te}$, as before. The instability criterion now requires that for some real ω and real k

$$\frac{\partial \Delta}{\partial \omega} = \frac{2\omega\omega_{pi}^2}{(\omega^2 - k^2 V_{Ti}^2)^2} + \frac{2\omega_{pb}^2}{(\omega - kv_0)^3} = 0 \qquad (3.41)$$

It is convenient to introduce the real variable $V = \omega/k$. Equation 3.41 can then be written as

$$\left(1 - \frac{V_{Ti}^2}{V^2}\right)^{-2} = \left(\frac{M}{m}\right)\left(\frac{n_b}{n_p}\right)\left(\frac{v_0}{V} - 1\right)^{-3} \qquad (3.42)$$

and Equation 3.40 as

$$\frac{\omega^2}{\omega_{pb}^2} = \left(\frac{v_0}{V} - 1\right)^{-2} + \left(\frac{m}{M}\right)\left(\frac{n_p}{n_b}\right)\left(1 - \frac{V_{Ti}^2}{V^2}\right)^{-1} - \left(\frac{n_p}{n_b}\right)\left(\frac{V^2}{V_{Te}^2}\right)$$

$$(3.43)$$

The criterion for an instability can now be stated as follows: If Equation 3.42 yields a real value of V that when substituted into Equation 3.43 gives a real value of ω, then an instability is obtained. It is convenient to introduce the normalized parameter $x = v_0/V$. Equation 3.42 can then be written as

$$\frac{1}{(x-1)^3} = \frac{\left(\frac{m}{M}\right)\left(\frac{n_p}{n_b}\right)}{\left(1 - \frac{V_{Ti}^2}{v_0^2}x^2\right)^2} \tag{3.44}$$

Using Equation 3.44 in Equation 3.43, we find

$$\frac{\omega^2}{\omega_{pb}^2} = (x-1)^{-2}\left\{1 + \left[\left(\frac{m}{M}\right)\left(\frac{n_p}{n_b}\right)\right]^{\frac{1}{2}}(x-1)^{\frac{1}{2}} - \left(\frac{n_p}{n_b}\right)\left(\frac{2V_0}{T_e}\right)\left(1 - \frac{1}{x}\right)^2\right\} \tag{3.45}$$

Both sides of Equation 3.44 are sketched in Figure 3.5 for the case $V_{Ti} < v_0$. We see that a real value of x (or V) <u>must</u> be obtained for $1 < x < v_0/V_{Ti}$, $(V_{Ti} < V < v_0)$. An exact solution in this case is difficult; however, we can make the following statement: <u>If the parameter $(n_b/n_p)(T_e/2V_0)$ is greater than unity, then an instability is obtained for all</u> $V_{Ti} < v_0$.

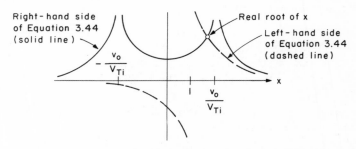

Figure 3.5. A real root of Equation 3.44.

The proof of this statement follows directly by noting that $\omega^2 > 0$ for any $x \geq 1$ in Equation 3.45 as long as $(n_b/n_p)(T_e/2V_0) \geq 1$, and a real root of $x > 1$ exists as long as $v_0 \geq V_{Ti}$, as noted before.

This condition is, of course, a sufficient condition for an instability and is not a necessary condition. It is an important statement, however, because it indicates that the meaningful quantities to be compared are the <u>velocities</u> of the beam and the ions (v_0 and V_{Ti}), and not the <u>energies</u> (V_0 and T_i). This statement is plausible from the wave picture of the interaction; a particular ion can be considered as cold as long as its unperturbed velocity is much less than the phase velocity of the wave, which must be on the order of v_0 for interaction with the beam.

In conclusion, a reactive-medium type of interaction with the ions can be obtained for ion temperatures such that $T_i < (M/m)2V_0$, as long as the electrons are sufficiently warm, $T_e > (n_p/n_b)2V_0$. The next section demonstrates that this instability is convective, and shows that the amplification rate can be very large when $(n_b/n_p)(T_e/2V_0)$ is greater than unity.

3.5 Strong Reactive-Medium Amplification with Ions

The previous section considered the general conditions necessary for an instability with the ions in a hot-electron plasma without regard for the type of instability or the growth rates in time or space. It will be shown here that a very interesting transition takes place when the beam density exceeds a certain critical value. This transition can be anticipated by looking at $k \to \infty$ as $\omega \to \omega_{pi}$ in the dispersion equation (3.33):

$$1 - \frac{\omega_{pi}^2}{\omega^2} \simeq \left(\frac{1}{k^2}\right)\left(\frac{\omega_{pb}^2}{v_0^2} - \frac{\omega_{pe}^2}{V_{Te}^2}\right) \tag{3.46}$$

In order to make this transition clear, we shall assume initially that $T_i = 0$. For $\beta_{pb} < \beta_{De}$, where $\beta_{De} = \omega_{pe}/V_{Te}$ is the Debye wave number and $\beta_{pb} = \omega_{pb}/v_0$, complex k is obtained for $\omega < \omega_{pi}$ and real k for $\omega > \omega_{pi}$, as in the case of a plasma without the electron beam. For $\beta_{pb} > \beta_{De}$, however, the situation is reversed, and complex k is obtained below ω_{pi}; that is, a transition occurs when the parameter $(n_b/n_p)(T_e/2V_0)$ is equal to unity.

Computations have been made on the dispersion equation with a square distribution for the electrons and cold ions. This dispersion equation is

$$1 = \frac{\omega_{pi}^2}{\omega^2} + \frac{\omega_{pe}^2}{\omega^2 - k^2 V_{Te}^2} + \frac{\omega_{pb}^2}{(\omega - kv_0)^2} \tag{3.47}$$

Note, however, that in all cases where $v_0 \ll V_{Te}$, the following results are identical to what would be obtained from a Maxwellian distribution for the electrons.

In Figures 3.6 through 3.8, plots of the complex k values for real ω are presented for the case of protons (M/m = 1836) with $n_b/n_p = 10^{-2}$ and for various values of the parameter

$$\eta = \left(\frac{n_b}{n_p}\right)\left(\frac{T_e}{2V_0}\right) \tag{3.48}$$

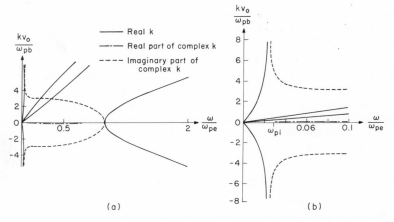

Figure 3.6. Dispersion for $\eta = 0.1$. (a) High frequency.
(b) Detail at low frequency.

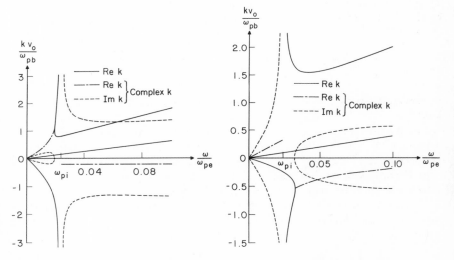

Figure 3.7. Dispersion for   Figure 3.8. Dispersion for
$\eta = 0.5$. $\eta = 4$.

For these parameters, the instability condition given by Inequality
3.37 requires that $\eta \geq 0.38$.

The case of $\eta = 0.1$ is presented in Figure 3.6. We see that no
point of $\partial\omega/\partial k = \infty$ for real ω and real k is obtained; that is, no
complex ω for real k results, in agreement with the instability
condition given by Inequality 3.37. The case of $\eta = 0.5$ is presented
in Figure 3.7, and, as expected, it indicates the presence of an instability. (Only the low-frequency region is shown in the remaining
figures because for $\omega \gg \omega_{pi}$ they are all qualitatively the same as

Figure 3.6a.) It seems physically clear, and will be demonstrated
in a moment for a different numerical case, that this instability is
convective and that the complex root of k below ω_{pi} with $k_i > 0$
represents the amplification rate. In Figure 3.8, the case of $\eta = 4$
is presented. As previously noted, the complex roots of k with
$k_i \to \infty$ at ω_{pi} now occur below ω_{pi}, and an intuitive interpretation
of Figure 3.8 indicates that this branch of the dispersion curve is
the amplification rate; that is, we now have an infinite amplification rate at $\omega = \omega_{pi}$.

The criteria of Chapter 2 have been applied to the case illustrated
in Figure 3.8 in order to prove these statements; the results are
shown in Figure 3.9. We see that there are no absolute instabilities
and that the complex root with $k_i > 0$ for real $\omega < \omega_{pi}$ is the amplifying wave, as expected.

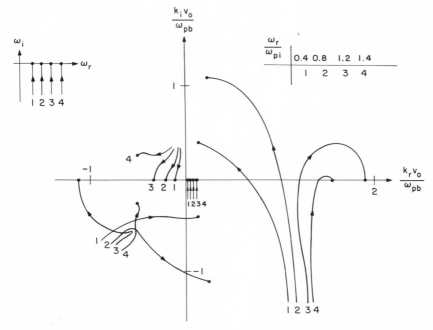

Figure 3.9. Amplification criteria for $\eta = 4$.

The dispersion for a very dense beam with $n_b/n_p = 1$ and for
protons is given in Figures 3.10 and 3.11 for the case of $T_e/2V_0$
equal to $\frac{1}{4}$ and 4. The phenomena are the same as before, the
only difference being the more extreme distortion of the beam and
plasma waves from their "unperturbed" shapes.

This appearance of infinite amplification rate at ω_{pi} could also
be anticipated by noting that as $T_e \to \infty$ the dispersion equation (3.47)
becomes the same as that of an electron beam moving through ions

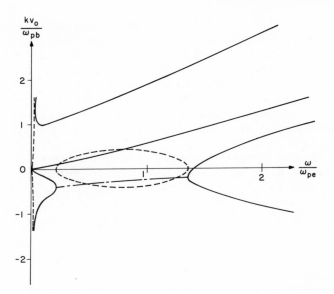

Figure 3.10. Dispersion for $n_b/n_p = 1$ and $\eta = \frac{1}{4}$.

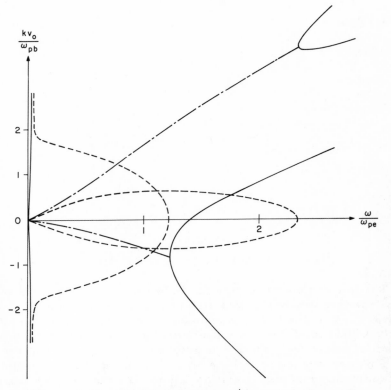

Figure 3.11. Dispersion for $n_b/n_p = 1$ and $\eta = 4$.

only. This system (for the assumed cold ions) has an infinite amplification rate, as is well known.[5] The condition that

$$\eta = \left(\frac{n_b}{n_p}\right)\left(\frac{T_e}{2V_0}\right) > 1 \tag{3.49}$$

can thus be interpreted as the critical electron temperature above which the plasma electrons cease to "short out" this beam-ion interaction.

The condition given by Equation 3.49 can also be written as $\lambda_{pb} < \lambda_{De}$, where $\lambda_{pb} = 2\pi v_0/\omega_{pb}$ is the beam space-charge wavelength and $\lambda_{De} = 2\pi V_{Te}/\omega_{pe}$ is the Debye wavelength in the plasma. This form is physically appealing because it suggests that a certain characteristic length associated with the beam, namely, the beam space-charge wavelength, should fit inside a Debye sphere in order to have a strong interaction with the ions.

3.5.1 Kinetic Power. It is interesting to consider the power balance in these complex waves near ω_{pi}. If the plasma were described by the macroscopic transport equations with an isotropic pressure, the dispersion equation would be the same as that given by Equation 3.47.[61] For this macroscopic model the small-signal power flow would be composed of the kinetic power transported by the electron beam[19] and the acoustic power carried by the plasma electrons.[61] For a single wave with ω real and k complex, the total time-average power flow must vanish,[61] so there must be a balance between the kinetic power and the acoustic power. It is easily shown that the sign of the kinetic power for a single wave is determined by $-\mathrm{Re}(k - \omega/v_0)$.[2] If we expand Equation 3.47 around $\omega \simeq \omega_{pi}$, we find that

$$k \simeq \pm j\alpha\left[1 + \frac{2\beta_e\beta_{pb}^2}{k(\beta_{pb}^2 - \beta_{De}^2)}\right]^{\frac{1}{2}} \simeq \pm j\alpha + \beta_e\left(\frac{\beta_{pb}^2}{\beta_{pb}^2 - \beta_{De}^2}\right) \tag{3.50}$$

with

$$\alpha^2 = -\frac{\beta_{pb}^2 - \beta_{De}^2}{1 - \frac{\omega_{pi}^2}{\omega^2}} \tag{3.51}$$

in the limit as $\alpha \to \infty$. For $\beta_{pb} < \beta_{De}$, the kinetic power P_K is positive, and therefore the acoustic power P_A must be negative in the wave with complex k ($\omega > \omega_{pi}$). For $\beta_{pb} > \beta_{De}$, the situation is reversed, and now P_K is negative in the complex wave ($\omega < \omega_{pi}$). The power flows in the latter case are analogous to those in the traveling-wave-tube (TWT) amplifier, where now the acoustic power plays the role of the circuit power in the TWT case. The signal on the beam grows in amplitude at the expense of the dc beam energy, and therefore the small-signal kinetic power on the beam increases (negatively) with distance, and the acoustic power increases positively with distance. The dc beam energy is thereby converted into the ac energy of oscillations of the beam and plasma particles and the electric field. It is emphasized that the interpretations based on this power analysis assume that the signal "flows" in the +z-direction, and is not a proof that this wave is indeed amplifying.

3.5.2 Finite Ion Temperature. The preceding analysis has shown that for zero ion temperature the amplification rate can become infinite for electron temperatures above a certain critical value. In Section 3.4 it was shown that instability persists as long as $v_0 > V_{Ti}$, when finite ion temperature is considered. The present analysis is concerned with determining the maximum amplification rate for very small ion temperatures.

If we use square distributions for both ions and electrons, the dispersion equation becomes

$$1 - \frac{\omega_{pi}^2}{\omega^2 - k^2 V_{Ti}^2} - \frac{\omega_{pe}^2}{\omega^2 - k^2 V_{Te}^2} - \frac{\omega_{pb}^2}{(\omega - kv_0)^2} = 0 \qquad (3.52)$$

For very low V_{Ti}, near ω_{pi} we can assume that $|k| \gg \omega/v_0$ and $|k| \gg \omega/V_{Te}$ for the roots of interest. Equation 3.52 then becomes biquadratic and can be solved for k^2 in the form

$$k^2 = a \pm \sqrt{a^2 - c^2} \qquad (3.53)$$

where

$$a = \frac{1}{2}\left(\frac{\omega^2 - \omega_{pi}^2}{V_{Ti}^2} + \beta_{pb}^2 - \beta_{De}^2\right) \qquad (3.54)$$

and

$$c^2 = \left(\frac{\omega_{pi}^2}{V_{Ti}^2}\right)\left(\beta_{pb}^2 - \beta_{De}^2\right) \qquad (3.55)$$

and where we have assumed that $\omega \simeq \omega_{pi}$. The quantity a contains the important frequency dependence; the qualitative behavior of k as a function of a will now be determined. We note that when $a^2 < c^2$ we have

$$k^2 = a \pm j \sqrt{c^2 - a^2} \qquad (3.56)$$

and therefore

$$|k^2| = c \qquad (3.57)$$

independent of a . Therefore, the maximum k_i in this range occurs when a = -c, as shown in Figure 3.12. The regions where

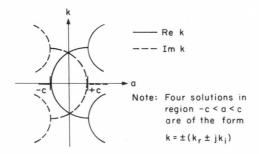

Figure 3.12. Dispersion near ω_{pi} for finite ion temperatures.

$a^2 > c^2$ are also shown in the figure. The larger k_i when a < -c corresponds to a cutoff "ion wave" and is not an amplifying wave because for $a^2 \gg c^2$ the roots of Equation 3.53 are

$$k^2 \simeq \frac{\omega^2 - \omega_{pi}^2}{V_{Ti}^2} \qquad (3.58)$$

and

$$k^2 \simeq \frac{\beta_{pb}^2 - \beta_{De}^2}{1 - \dfrac{\omega_{pi}^2}{\omega^2}} \qquad (3.59)$$

The maximum amplification rate is therefore obtained at a = -c and is equal to

$$(k_i)_{max} = \left(\beta_{pb} \frac{\omega_{pi}}{V_{Ti}} \right)^{\frac{1}{2}} \left(1 - \frac{\beta_{De}^2}{\beta_{pb}^2} \right)^{\frac{1}{4}} = \beta_{pb} \left[\left(\frac{n_p}{n_b} \right) \left(\frac{2V_0}{T_i} \right) \right]^{\frac{1}{4}} \left[1 - \left(\frac{n_p}{n_b} \right) \left(\frac{2V_0}{T_e} \right) \right]^{\frac{1}{4}}$$

$$(3.60)$$

We see that the amplification rate can be on the order of β_{pb} even when T_i is on the order of $(n_p/n_b)2V_0$. Once again the analysis has indicated that this interaction should persist over a wide range of ion temperatures as long as the plasma electrons are sufficiently warm.

3.5.3 Relativistic Temperatures. For practical electron beams generated in the laboratory, the attainment of the condition (n_b/n_p) $(T_e/2V_0) > 1$ would probably require electron temperatures in the relativistic range. In this subsection the extension of this condition to the relativistic case will be given.

The dielectric constant for longitudinal waves in a relativistic plasma has been given by a number of authors. For our purpose, only the expansion of $K_{\parallel}(\omega, k)$ in the limit of $k \to \infty$ is required; this is simply*

$$K_{\parallel}(\omega, k) = 1 + \frac{\omega_{pe}^2}{k^2 \left(\frac{eT}{m}\right)} \tag{3.61}$$

for the Maxwell-Boltzman (Jüttner) distribution given by

$$f_0 = \left(\frac{1}{4\pi c^3}\right)\left(\frac{B}{K_2(B)}\right)e^{-B\gamma} \tag{3.62}$$

where

$$B = \frac{m_0 c^2}{eT} \tag{3.63}$$

$$\gamma = \left(1 - \frac{v^2}{c^2}\right)^{-\frac{1}{2}} \tag{3.64}$$

and ω_{pe}^2 is defined with the rest mass m_0. The condition for a "flip" at the pole at $\omega = \omega_{pi}$ is now

$$\frac{\omega_{pb,\ell}^2}{v_0^2} > \frac{\omega_{pe}^2}{\dfrac{eT}{m}} \tag{3.65}$$

where

---

*See, for example, Equations 23 and 24 in Imre.[63]

$$\omega_{pb,\ell}^2 = \frac{e\rho_{0b}}{\epsilon_0 m_\ell} \tag{3.66}$$

and the "longitudinal mass" of the beam electrons is defined as

$$m_\ell = m_0 \left(1 - \frac{v_0^2}{c^2}\right)^{-\frac{3}{2}} \tag{3.67}$$

Therefore, the condition for infinite gain at $\omega = \omega_{pi}$ is that

$$\left(\frac{n_b}{n_p} \frac{eT_e}{m_0}\right) \frac{\left(1 - \frac{v_0^2}{c^2}\right)^{\frac{3}{2}}}{v_0^2} \geq 1 \tag{3.68}$$

We see that the condition given by Equation 3.49 is not altered if the beam is nonrelativistic but the plasma has a relativistic temperature. For a relativistic beam, Equation 3.68 can be written in the form

$$\left(\frac{n_b}{n_p} \frac{T_e}{2V_0}\right) \geq \left(1 + \frac{V_0}{V_R}\right)\left(1 + \frac{1}{2}\frac{V_0}{V_R}\right) \tag{3.69}$$

where $V_R = m_0 c^2/e \simeq 0.5 \times 10^6$ volts, the relativistic unit of energy in electron volts. Equation 3.69 follows from Equation 3.68 by using the relation

$$\frac{v_0^2}{c^2} = 1 - \frac{1}{\left(1 + \frac{V_0}{V_R}\right)^2} \tag{3.70}$$

In the case of a relativistic beam, the condition is more difficult to achieve.

B. TRANSVERSE INTERACTIONS

3.6 Dispersion Equation for Transverse Waves

The dispersion equation for transverse waves in an electron-beam plasma system is derived in Appendix C. If we assume that the electron beam is cold and that the plasma particles have symmetrical distribution functions, the dispersion equation can be written as[26]

$$\frac{k^2 c^2}{\omega^2} = 1 - \sum \frac{\omega_p^2}{\omega} \int \frac{f_0(\overline{v})\ d^3\overline{v}}{\omega - kv_z \mp \omega_c} - \frac{\omega_{pb}^2(\omega - kv_0)}{\omega^2(\omega - kv_0 \pm \omega_{ce})} \qquad (3.71)$$

The sum is over all species of the plasma, ω_c carries the sign of
the charge, and the upper sign applies to left-handed polarization,
the lower to right-handed polarization. For the left-polarized
wave the fields are related as

$$\frac{E_x}{E_y} = -j \qquad (3.72)$$

that is, the fields rotate in time in the sense of gyration of a positively charged particle about the steady magnetic field.*
    As in the longitudinal case, the integration over $\overline{v}$ in Equation
3.71 defines two functions of ω and k, depending on whether
$\text{Im}\,(\omega \mp \omega_c)/k$ is assumed to be positive or negative when the integration is carried out. In all of Part B only the form obtained
for $\text{Im}\,(\omega \mp \omega_c)/k < 0$ will be given because it turns out that this
includes all the unstable wave solutions.
    As in the longitudinal case, it is helpful to work initially with
distribution functions that are mathematically simpler than the
Maxwellian distribution, in order to obtain some feeling for the
results to be expected. In the following sections, extensive use
will be made of the resonance distribution, which in three-dimen-
sional form is

$$f_0(\overline{v}) = \frac{V_T}{\pi^2} \frac{1}{(v^2 + V_T^2)^2} \qquad (3.73)$$

where $v^2 = v_x^2 + v_y^2 + v_z^2$. Integration over the transverse velocities gives

$$f_0(v_z) = \int_0^\infty f_0(\overline{v}) 2\pi v_t\ dv_t = \left(\frac{V_T}{\pi}\right)\left(\frac{1}{v_z^2 + V_T^2}\right) \qquad (3.74)$$

which is the same as was used in Section 3.2. The evaluation of
the integration in the dispersion equation (3.71) follows most easily
by closing the contour in the upper half of the complex v_z-plane,
obtaining just the residue at the simple pole $v_z = jV_T$, for

---

*One should note that the designations of right and left polariza-
tion are reversed if the frequency is negative.

$\text{Im}\,(\omega \mp \omega_c)/k < 0$, so that the pole at $v_z = (\omega \mp \omega_c)/k$ lies below the contour. The result is

$$\frac{k^2 c^2}{\omega^2} = 1 - \sum \frac{\omega_p^2}{\omega(\omega \mp \omega_c - jkV_T)} - \frac{\omega_{pb}^2(\omega - kv_0)}{\omega^2(\omega - kv_0 \pm \omega_{ce})} \qquad (3.75)$$

The solution of Equation 3.75, in the absence of the beam, shows that the transverse waves in a collisionless plasma are damped by a process that is analogous to the Landau damping of longitudinal waves. This transverse damping is commonly called "cyclotron damping."[26] For a cold plasma in the absence of the beam, the dispersion equation is

$$\frac{k^2 c^2}{\omega^2} = 1 - \frac{\omega_{pi}^2}{\omega(\omega \mp \omega_{ci})} - \frac{\omega_{pe}^2}{\omega(\omega \pm \omega_{ce})} \qquad (3.76)$$

This dispersion is sketched in Figure 3.13. As would be expected physically, the left-polarized wave has a resonance at the ion cyclotron frequency, since it is polarized in the sense of rotation of

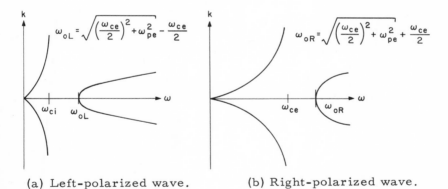

(a) Left-polarized wave.   (b) Right-polarized wave.

Figure 3.13. Transverse waves in a cold plasma.

the ions. Conversely, the right-polarized wave has a resonance at the electron cyclotron frequency. The cyclotron damping becomes important near these resonances even for a very "cool" plasma, as can be seen from Equation 3.75.

3.7 Transverse Waves on Electron Beams and Weak-Coupling Predictions

In order to understand the transverse-wave instabilities in a beam-plasma system, it is helpful to determine first the types of propagating waves that can exist on the beam in the absence of the plasma.   An application of the formalism of weak coupling, familiar from microwave beam tube theory,[19,48,49] then allows one to make intuitive predictions about the appearance of an instability.   In this approach, an instability results when a beam wave that carries negative small-signal energy (an active wave) is in synchronism with a propagating plasma wave, that is, when the two waves have ω and k approximately the same.   It should be emphasized that the application of this weak-coupling principle is not a rigorous proof, since the beam is very strongly coupled to the plasma; however, the exact analysis given in the following sections agrees well with these weak-coupling predictions.

   3.7.1   Electron-Beam Waves.   The dispersion equation for the transverse waves on a cold electron beam follows directly from Equation 3.71 by omitting the plasma terms:

$$k^2 c^2 = \omega^2 - \frac{\omega_{pb}^2 (\omega - k v_0)}{\omega - k v_0 \pm \omega_{ce}} \tag{3.77}$$

For a relativistic beam, the relativistic <u>transverse</u> mass should be used in computing ω_{pb} and ω_{ce}; that is,

$$\omega_{pb}^2 = \frac{e^2 n_b}{\epsilon_0 m_t} \tag{3.78}$$

$$\omega_{ce} = \frac{e B_0}{m_t} \tag{3.79}$$

where

$$m_t = \gamma_0 m_0 \tag{3.80}$$

$$\gamma_0 = \left(1 - \frac{v_0^2}{c^2} \right)^{-\frac{1}{2}} \tag{3.81}$$

and m_0 is the rest mass of an electron.

   In Appendix F the qualitative nature of the dispersion of these transverse beam waves is determined by transforming into a reference frame moving with the beam.   This dispersion is sketched in Figure 3.14 for the case of a relatively low density beam ($\omega_{pb} \ll \omega_{ce}$).   In Appendix F it is shown that the only active waves are the branches of the left-polarized wave with phase velocity less than the beam velocity v_0.   There are two such branches: the "slow cyclotron wave" with $k \simeq (\omega + \omega_{ce})/v_0$, and the low-frequency region that is shown in more detail in Figure 3.15.   Note that a portion of

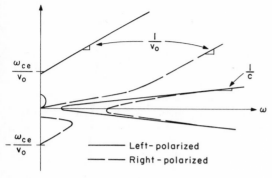

Figure 3.14. Beam dispersion.

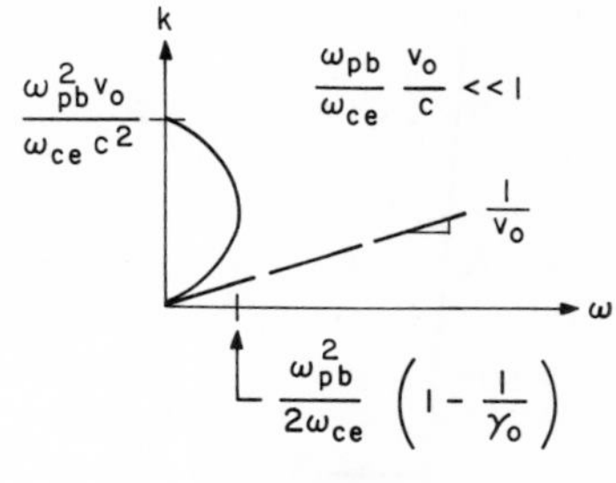

Figure 3.15. Detail of low-frequency backward wave on the beam.

this low-frequency branch has $\partial\omega/\partial k < 0$, which means that it carries negative small-signal energy in the direction <u>opposite</u> to the beam velocity.

3.7.2 Weak-Coupling Predictions. The theory of coupling of modes would lead one to expect an instability when an electron-beam wave that carries negative small-signal energy is in synchronism with a plasma wave. In Figure 3.16 the dispersion for the left-polarized plasma wave for a cold plasma is sketched along

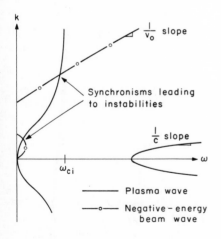

Figure 3.16. Superimposed left-polarized waves of plasma and electron beam.

with the negative-energy beam waves, under the assumption of a relatively weak beam $[(\omega_{pb}/\omega_{ce})(v_0/c) \ll 1]$. Note carefully that since the total beam-plasma system must have zero net charge, the ion density must exceed the electron density in the plasma by an amount just equal to the beam density.[*] It is easily shown that

---

[*]See Section 3.9.

in this case ω/k for the plasma wave tends to zero (rather than the Alfvén speed) as ω tends to zero, as is indicated in Figure 3.16.

From coupling of modes we would expect a convective instability to occur at the intersection near ω_{ci}, and an absolute instability to occur in the low-frequency regime. The following sections verify that these instabilities are predicted by an analysis of the exact dispersion equation, and consider the growth rates as well as the effects of a finite temperature. Some doubts are expressed in Section 3.9, however, about the validity of the one-dimensional model in the regime where the very low frequency instability occurs.

3.8 Interaction at the Ion Cyclotron Frequency

In the previous section it was shown that a heuristic application of coupling of modes would lead one to expect a convective instability of the transverse wave near the ion cyclotron frequency in a cold plasma. In this section this instability will be considered in some detail. It will be shown that under most conditions the instability is greatly altered by a finite ion temperature, because of the cyclotron damping, and that this damping can itself lead to a resistive-medium type of instability.

3.8.1 Cold Plasma. The dispersion equation for a cold plasma is obtained by setting $f_0(\overline{v}) = \delta(\overline{v})$ in Equation 3.71:

$$\omega^2 - k^2 c^2 - \frac{\omega_{pe}^2 \omega}{\omega \pm \omega_{ce}} - \frac{\omega_{pi}^2 \omega}{\omega \mp \omega_{ci}} = \frac{\omega_{pb}^2(\omega - kv_0)}{\omega - kv_0 \pm \omega_{ce}} \qquad (3.82)$$

where again the upper sign is for the left-polarized wave and the lower sign is for the right-polarized wave. Equation 3.82 is of the form

$$\Delta(\omega, k) = \frac{\omega_{pb}^2(\omega - kv_0)}{\omega - kv_0 \pm \omega_{ce}} \qquad (3.83)$$

where the dispersion equation for the plasma waves is $\Delta(\omega, k) = 0$. The dispersion equation for the complete beam-plasma system is cubic in the wave number k and fifth order in the frequency ω. For a relatively weak beam (that is, a low-density beam), some of the solutions of Equation 3.83 are given by $\Delta(\omega, k) \simeq 0$ unless $kv_0 \simeq \omega \pm \omega_{ce}$.[*] In other words, the form of Equation 3.83 shows that the plasma waves are appreciably perturbed by such a weak

---

[*]Or unless ω and k both tend to zero, as considered in the next section.

beam only if the wave is in "resonance" with the beam electrons, that is, only if $\omega - kv_0 \simeq \mp \omega_{ce}$, so that in a reference frame moving with the beam the field rotates in the right-handed sense at the electron cyclotron frequency. (Note that a negative frequency reverses the left and right polarization "labels.")

We shall attempt a solution for the beam waves by writing $k = [\omega \pm \omega_{ce})/v_0] + \delta$, and assuming $|\delta| \ll (\omega \pm \omega_{ce})/v_0$. As in Section 3.3, δ is essentially the "correction" in $k(\omega)$ at some real ω arising from the (small) beam density. It is determined from

$$\pm \frac{\omega_{pb}^2 \omega_{ce}}{\delta v_0} \simeq \Delta\left(k = \frac{\omega \pm \omega_{ce}}{v_0}\right) + \delta\left(\frac{\partial \Delta}{\partial k}\right)_{k=\frac{\omega \pm \omega_{ce}}{v_0}} \quad (3.84)$$

In Figure 3.17 the resulting real k - real ω diagram is drawn for such a weak beam for both left-polarized waves (upper sign) and right-polarized waves (lower sign), neglecting the region near ω and k tending to zero, which is considered in the next section. This diagram was constructed by using Equation 3.84 to find the perturbations in k near the intersection of the plasma waves with the "beam waves," $k = (\omega \pm \omega_{ce})/v_0$. The dispersion near these intersections has exactly the form one would expect from the coupling of modes formalism; the discussion in Section 2.7 allows us to assert immediately that the complex k with $k_i > 0$ for the left-polarized wave is amplifying, while in the right-polarized case it represents only evanescent waves.

The maximum amplification rate of the left-polarized wave can be determined from Equation 3.84. It occurs at the frequency for which

$$\Delta\left(k = \frac{\omega + \omega_{ce}}{v_0}\right) = 0 \quad (3.85)$$

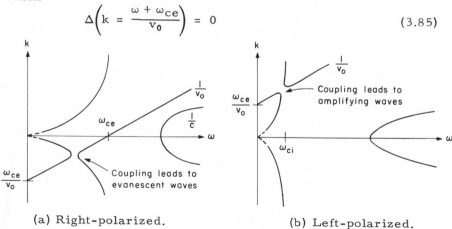

(a) Right-polarized.   (b) Left-polarized.

Figure 3.17. Real ω - k diagram for transverse waves in a beam-plasma system.

and since

$$\frac{\partial \Delta}{\partial k} = -2kc \tag{3.86}$$

we find

$$(\text{Im } \delta)_{max} \simeq \frac{\omega_{pb}}{\sqrt{2} c} \tag{3.87}$$

Equation 3.87 is clearly the maximum $|\delta|$ for any frequency, and therefore for the weak-beam approximation ($|\delta| \ll \omega_{ce}/v_0$) to be valid, we must have

$$\frac{\omega_{pb}}{\omega_{ce}} \frac{v_0}{c} \ll 1 \tag{3.88}$$

In almost all cases of practical interest, this inequality is well satisfied.

3.8.2 Warm Plasma. It can be shown that the electron temperature of the plasma has essentially no effect on this interaction, as would be expected physically because the electron motion is negligible near ω_{ci}. However, for the ion temperature to be neglected, it is necessary that $|\omega - \omega_{ci}| \gg kV_{Ti}$ near the intersection of the plasma wave and the beam cyclotron wave (see Equation 3.71). Solving for $\omega_{ci} - \omega$ at $k \simeq \omega_{ce}/v_0$ from Equations 3.85 and 3.83, we obtain

$$\omega_{ci} - \omega \simeq \left(\frac{v_0^2}{c^2}\right)\left(\frac{\omega_{pi}^2}{\omega_{ce}^2}\right)\omega_{ci} \tag{3.89}$$

Therefore, the condition for neglecting the ion temperature in this interaction is that

$$\frac{V_{Ti}}{v_0} \ll \left(\frac{m}{M}\right)\left(\frac{v_0^2}{c^2}\right)\left(\frac{\omega_{pi}^2}{\omega_{ce}^2}\right) \tag{3.90}$$

or that[*]

$$\frac{T_i}{2V_0} \ll \left(\frac{m}{M}\right)\left(\frac{v_0}{c}\right)^4\left(\frac{\omega_{pi}}{\omega_{ce}}\right)^4 \tag{3.91}$$

---

[*]A similar condition has been derived by Stix, page 210 of Reference 26.

As a numerical example, consider protons with density $10^{12}/cc$, magnetic field of 1000 gauss, and $v_0/c = 0.2$; Inequality 3.91 requires that the ion temperature be much less than 10^{-7} volt, or much less than $1°$ K! Even for a magnetic field of 100 gauss, this still requires ion temperatures less than 10^{-3} volt. We conclude that for most cases of practical interest the ion temperature cannot be neglected in the analysis of this interaction.

The inclusion of the ion temperature brings in the cyclotron damping of the plasma waves near the resonance at $\omega \simeq \omega_{ci}$. Since the slow cyclotron wave on the beam carries negative small-signal energy, one might anticipate that the "lossyness" of the plasma medium could result in a resistive-medium type of interaction for this wave. To investigate this, we shall first consider the dispersion equation obtained when resonance-type distribution functions are used for the plasma ions and electrons (Equation 3.75). It can be written in the form

$$\frac{\omega_{pb}^2(\omega - kv_0)}{\omega - kv_0 \pm \omega_{ce}} = \Delta(\omega, k) \qquad (3.92)$$

where now

$$\Delta(\omega, k) = \omega^2 - k^2c^2 - \frac{\omega_{pi}^2\omega}{\omega \mp \omega_{ci} - jkV_{Ti}} - \frac{\omega_{pe}^2\omega}{\omega \pm \omega_{ce} - jkV_{Te}}$$

$$(3.93)$$

for positive k_r, for which the $\Delta_1(\omega, k)$ dispersion equation described in Appendix B is appropriate. For negative k_r, the sign of the jkV_T term is changed. As before, we assume a relatively weak beam so that the solution for the beam waves can be determined by setting $k = [(\omega \pm \omega_{ce})/v_0] + \delta$ and by assuming $\delta \ll (\omega \pm \omega_{ce})/v_0$. We then find

$$\pm \frac{\omega_{pb}^2\omega_{ce}}{v_0} \simeq \Delta\left(k = \frac{\omega \pm \omega_{ce}}{v_0}\right) \qquad (3.94)$$

From Equation 3.93, we see that Im $\Delta < 0$ for real ω and real k (either positive or negative); from Equation 3.94 it follows that Im $\delta > 0$ for the left-polarized wave, while Im $\delta < 0$ for the right-polarized wave. Therefore, if we fix k as real, since Im $\omega = -$Im (δv_0), we see that the left-polarized wave is unstable, while the right-polarized wave is damped. The instability of the left-polarized wave is clearly convective (since the dispersion equation is first order in δ), and Im δ is the amplification rate that represents growth in the direction of the beam flow.

This amplification rate is of reasonably large amplitude only when Δ has a significant imaginary part, a condition that occurs near $\omega \simeq \omega_{ci}$. For $k \simeq \omega_{ce}/v_0$, the k^2c^2 term in Equation 3.93 is much larger than the term arising from the ions if the inequality in 3.90 is reversed, that is, if

$$\frac{V_{Ti}}{v_0} \gg \left(\frac{m}{M}\right)\left(\frac{v_0^2}{c^2}\right)\left(\frac{\omega_{pi}^2}{\omega_{ce}^2}\right) \tag{3.95}$$

which, as pointed out before, is well satisfied even for relatively "cool" ions in most cases of interest. It is then clear that the maximum Im δ occurs at $\omega = \omega_{ci}$. From Equations 3.93 and 3.94, we find, using Inequality 3.95, that at $\omega = \omega_{ci}$

$$\delta \simeq -\left(\frac{\omega_{pb}^2}{\omega_{ce}}\right)\left(\frac{v_0}{c^2}\right)\left[1 - \left(j\frac{m}{M}\right)\left(\frac{\omega_{pi}^2}{\omega_{ce}^2}\right)\left(\frac{v_0^2}{c^2}\right)\left(\frac{v_0}{V_{Ti}}\right)\right] \tag{3.96}$$

and therefore

$$(\text{Im }\delta)_{max} = \left(\frac{\omega_{ci}}{V_{Ti}}\right)\left(\frac{\omega_{pi}^2\omega_{pb}^2}{\omega_{ce}^4}\right)\left(\frac{v_0^4}{c^4}\right) \tag{3.97}$$

In the preceding derivation, both the term arising from the electrons and the ω^2 term in Equation 3.93 are ignored, since they are usually much less than $k^2c^2 \simeq \omega_{ce}^2(c^2/v_0^2)$. From Equation 3.96, we see that the validity of the weak-beam assumption again requires that the inequality given by Equation 3.88 be satisfied.

The amplification rate of this resistive-medium type of interaction (Equation 3.97) is generally quite small. As a numerical example, for protons of density $10^{12}/cc$, magnetic field of 1000 gauss, $v_0/c = 0.2$, beam density of $10^{10}/cc$, and ions of 0.1 eV temperature, this growth rate is on the order of 10^{-5} cm^{-1}, which is a very weak instability.

The dispersion equation resulting from the Maxwellian distribution (Equation 3.5) is essentially the same as the resonance distribution except in the regime $|\omega - \omega_{ci}| \gg kV_{Ti}$. (Compare with the analogous discussion for the longitudinal case in Section 3.2.) For this regime the Im $\Delta(\omega, k)$ is exponentially small with $[(\omega - \omega_{ci})/kV_{Ti}]^2$ for real ω and real k.[26] Therefore, the growth rate of the resistive-medium amplification is much more peaked around $\omega \simeq \omega_c$ for a Maxwellian than for a resonance distribution. The maximum growth rate is still at $\omega = \omega_{ci}$, however, since the largest Im $\Delta(\omega, k)$ for real ω and k occurs here.[26] To determine this peak growth rate, we must evaluate the integral in Equation 3.71 at $\omega = \omega_{ci}$; this yields

$$\int_{-\infty}^{+\infty} \frac{f_{0i}(v_z)\, dv_z}{kv_z} = -j\pi\left(\frac{1}{k}\right) f_{0i}(v_z = 0) \tag{3.98}$$

for $k_r > 0$ and any $f_0(v_z)$ that is an even function of v_z. Equation 3.98 gives $-j\sqrt{(\pi/2)}/kv_{Ti}$ for a Maxwellian distribution and $-j/kV_{Ti}$ for the resonance distribution used before. Therefore, the maximum amplification given by Equation 3.97 is merely multiplied by $\sqrt{\pi/2}$ if a Maxwellian distribution is used.

3.9 Alfvén Wave Instability

In Section 3.7 it was shown that there is an active backward wave on a one-dimensional electron beam which should result in an absolute instability when coupled with the forward-traveling Alfvén wave in the plasma. In this section it is shown that an instability is also predicted by an analysis of the exact dispersion equation.

To analyze the low-frequency regime, we assume a cold plasma (which is usually a good approximation in this regime) and expand the dispersion equation (3.82) under the assumption that $\omega \ll \omega_{ci}$ and $kv_0 \ll \omega_{ce}$. The condition of charge neutrality turns out to be extremely crucial in this regime, even in the limit as the beam density approaches zero. That is, we must have

$$n_b + n_{pe} = n_{pi} \tag{3.99}$$

for charge neutrality (for multiply charged ions, n_{pi} should be multiplied by that integer). Expanding Equation 3.82, we obtain

$$k^2 c^2 = \omega^2 + \left(\frac{\omega_{pe}^2}{\omega_{ce}\omega_{ci}}\right)\omega^2 + \left(\frac{\omega_{pb}^2}{\omega_{ce}}\right)\omega - \left(\frac{\omega_{pb}^2}{\omega_{ce}}\right)(\omega - kv_0) \tag{3.100}$$

for the left-polarized wave, and where we have used the fact that

$$\frac{\omega_{pb}^2}{\omega_{ce}} = \frac{\omega_{pi}^2}{\omega_{ci}} - \frac{\omega_{pe}^2}{\omega_{ce}} \tag{3.101}$$

If we define

$$k_0 = \frac{\omega_{pb}^2 v_0}{\omega_{ce} c^2} = \frac{\rho_{ob} v_0}{\left(\dfrac{B_0}{\mu_0}\right)} = \frac{J_{ob}}{\left(\dfrac{B_0}{\mu_0}\right)} \tag{3.102}$$

and[61]

$$\frac{c^2}{u_\alpha^2} = \frac{\omega_{pe}^2}{\omega_{ce}\omega_{ci}} \tag{3.103}$$

where u_α is the Alfvén speed in the plasma, Equation 3.100 can be written in the form

$$\omega^2 = k(k - k_0)u_\alpha^2 \tag{3.104}$$

for $u_\alpha \ll c$.

This dispersion equation is illustrated in Figure 3.18. There are complex roots of ω for real k between zero and k_0; for real ω, k is always real.

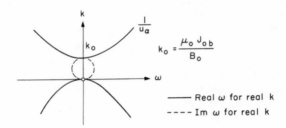

Figure 3.18. Low-frequency instability.

We note that the original assumption of $kv_0 \ll \omega_{ce}$ requires that $(\omega_{pb}^2/\omega_{ce}^2)(v_0^2/c^2) \ll 1$ in order that Equation 3.104 be valid. (Compare with the weak-beam assumption in Section 3.8.1, Equation 3.88.) However, even if this condition is not satisfied, there is still an instability. This follows from the fact that the dispersion equation (3.104) is rigorously correct in the limit of $k \to 0$ and $\omega \to 0$ where there are always complex roots of ω for real k.

It is easily shown that the dispersion equation (3.104) describes an absolute instability. For sufficiently large negative imaginary part of ω, the two roots of k are in different halves of the complex k-plane; these roots merge into a double root at the frequency

$$\omega_s = -j\,\frac{k_0 u_\alpha}{2} \tag{3.105}$$

and at the real wave number

$$k_s = \frac{1}{2}k_0 \tag{3.106}$$

The spatial pattern of this absolute instability is pure harmonic, whereas the growth in time is purely exponential with no oscillation

An analysis of the right-polarized wave shows that a similar absolute instability is obtained, only with $-k_0$ appearing in Equation 3.104 instead of k_0. This represents exactly the same physical phenomenon, since a reversal in the sign of <u>both</u> ω_r and k_r merely reverses the labeling of right- and left-polarized waves. Note that the <u>spatial</u> pattern in both cases corresponds to a right-handed helix about the magnetic field and that the vectors ($\overline{E}, \overline{H}$, and so on) at a fixed point in space remain in the same direction and increase exponentially with time.

We should remark in closing, however, that there is a certain lack of consistency in our model in the regime where this instability occurs, and this casts some doubts on the preceding description of the instability. We have, throughout this chapter, ignored the steady magnetic field created by the dc beam current J_{ob}. Normally, this oversight is excusable because this magnetic field is usually much smaller than the applied magnetic field B_0. For a beam of radius R, the maximum self-magnetic field is

$$H_{ob} = \frac{J_{ob}R}{2} \tag{3.107}$$

For the applicability of the one-dimensional model, however, the wavelength should be much smaller than the dimensions of the system, which requires

$$kR \gg 1 \tag{3.108}$$

The largest k value for this instability to be obtained is $k_0 = J_{ob}/H_0$, and for the one-dimensional assumption to still be valid in this regime, we must have $H_{ob} \gg H_0$ according to Equation 3.107 and Inequality 3.108.

We could eliminate this difficulty by introducing a compensating current such that the total dc current is zero; for instance, by considering a stream composed of an equal density of electrons and ions. This is certainly a different physical situation from the unneutralized stream; however, if one analyzes the new model, one finds an <u>entirely different</u> dispersion equation for this Alfvén wave instability[21,25] (which does, however, predict instability under the appropriate circumstances). None of the other instabilities analyzed in this chapter suffer from such an extreme sensitivity to the compensation (or lack of compensation) in the dc beam current.

Chapter 4

INTERACTIONS WITH A COLD PLASMA IN SYSTEMS OF FINITE TRANSVERSE DIMENSIONS

The results that were derived in the previous chapter for a one-dimensional beam-plasma system are strictly valid only when the wavelength is much smaller than the dimensions of either the electron beam or the plasma, as was mentioned in the introduction to Chapter 3. Since this is almost never true for laboratory-scale phenomena, solutions to the beam-plasma interaction problem that take account of the finite transverse dimensions are of utmost importance.

Unfortunately, the rigorous analysis of such finite systems is exceedingly complicated, even for cases in which the effects of temperature are ignored. In this investigation a number of simplifying assumptions will be made in order to obtain tractable results. For example, almost all of the results are derived within the so-called "quasi-static" approximation.[1,64] This means that we restrict ourselves to the study of waves with phase velocity much less than the speed of light, and to systems with transverse dimensions much less than a free-space wavelength.[*] In addition, a number of other approximations are made in the different sections, such as very thin beams and/or very weak beams. These limiting cases were chosen both because they render the mathematics more tractable and because they correspond to (somewhat) reasonable idealizations of a number of practical systems.

In Section 4.1 the simplest situation is analyzed; that of a very low density electron beam and a cold plasma, both filling a cylindrical waveguide structure. It is shown that there are a large variety of interactions at both high and low frequencies; these various interactions are catalogued and the amplification rates and starting lengths for oscillation are determined. In Section 4.2 the case of a very thin solid electron beam in a plasma that fills a circular waveguide is considered. Both the circularly symmetric

---

[*]A more precise statement of the requirements necessary for the validity of a quasi-static assumption in the analysis of waves in a cold plasma are given in References 61 and 65. Roughly, it requires that pc be much greater than ω and/or ω_{p0} over most frequencies, where p is the radial wave number. An outstanding exception to this rule is the band of frequencies near plasma resonance where quasi-statics becomes exact for any size plasma.

modes (n = 0) and the modes of one azimuthal variation (n = 1) are analyzed. Finally, Section 4.3 illustrates some interesting effects that can occur for very dense electron beams in the case of an infinite steady magnetic field. Specifically, it is shown that there is a transition from finite to infinite amplification rate at the electron plasma frequency for a critical value of the beam perveance.[28] This critical perveance is determined for both solid beams and hollow beams.

4.1 Interactions in a Waveguide Filled with a Weak Beam and a Plasma

In this section the interaction of a very weak beam with a cold plasma will be analyzed for the case in which both beam and plasma are homogeneous and fill a cylindrical waveguide structure (Figure 4.1). There is a steady magnetic field $\overline{B}_0$ aligned with the

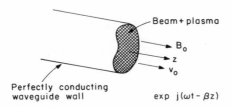

Figure 4.1. Beam-plasma system.

axis of the guide (z-direction). It is assumed that a quasi-static approximation is valid, so that the electric field vector can be written as the gradient of a scalar potential; that is,

$$\overline{E} = -\nabla\varphi \tag{4.1}$$

The dispersion equation for this system is derived in Appendix G, where it is shown that the potential φ must satisfy the two-dimensional Helmholtz equation

$$\nabla_T^2\varphi + p^2\varphi = 0 \tag{4.2}$$

where ∇_T^2 is the Laplacian operator in the transverse plane and all quantities are assumed to vary as $\exp[j(\omega t - \beta z)]$. The transverse wave number (or "transverse eigenvalue") is given by

$$p^2 = -\beta^2\left(\frac{K_\parallel^0}{K_\perp^0}\right) \tag{4.3}$$

where

$$K_{\parallel}^{0} = K_{\parallel} - \frac{\omega_{pb}^2}{(\omega - \beta v_0)^2} \qquad (4.4)$$

$$K_{\perp}^{0} = K_{\perp} - \frac{\omega_{pb}^2}{(\omega - \beta v_0)^2 - \omega_{ce}^2} \qquad (4.5)$$

and[61]

$$K_{\parallel} = 1 - \frac{\omega_{pi}^2}{\omega^2} - \frac{\omega_{pe}^2}{\omega^2} = 1 - \frac{\omega_{po}^2}{\omega^2} \qquad (4.6)$$

$$K_{\perp} = 1 - \frac{\omega_{pi}^2}{\omega^2 - \omega_{ci}^2} - \frac{\omega_{pe}^2}{\omega^2 - \omega_{ce}^2} = 1 - \frac{\omega_{po}^2(\omega^2 - \omega_{ce}\omega_{ci})}{(\omega^2 - \omega_{ce}^2)(\omega^2 - \omega_{ci}^2)} \qquad (4.7)$$

The boundary condition on the electric field requires that $\varphi = 0$ on the surface of the perfectly conducting waveguide wall. This determines an infinite set of eigenvalues for p^2, which depend only on the geometry of the waveguide and are positive real constants. For example, in a waveguide of circular cross section of radius b the solution of Equation 4.2 is of the form

$$\varphi(r, \phi) = A J_n(pr) e^{jn\phi} \qquad (4.8)$$

and the boundary condition on φ requires that

$$pb = \epsilon_{nm} \qquad (4.9)$$

where ϵ_{nm} is the mth zero of $J_n(x)$.

The dispersion of the plasma waves in the absence of the beam is sketched in Figure 4.2. As in the one-dimensional case, for a very weak beam $\omega_{pb} \to 0$, the plasma waves are relatively unperturbed unless they are in "synchronism" with the beam. By inspection of the dispersion equation, we see that such perturbations in the plasma waves occur when

$$\omega - \beta v_0 \simeq \pm \omega_{ce} \qquad \qquad (4.10)$$

or when

$$\omega - \beta v_0 \simeq 0 \qquad \qquad (4.11)$$

The first condition (Equation 4.10) clearly corresponds to a "resonance" of the plasma wave with the transverse motion of the

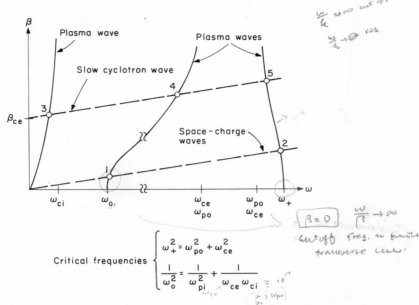

Figure 4.2. Plasma dispersion with coupling regions indicated.

beam electrons, whereas the second (Equation 4.11) corresponds
to a "resonance" with the longitudinal motion. Stated in another
way, the plasma waves are significantly perturbed only when they
are synchronous with the fast and slow space-charge waves or with
the fast and slow cyclotron waves on the beam.*

From the theory of coupling of modes,[48,49] we expect that only
the synchronisms with the active beam waves should lead to amplification or instability. The active beam waves are the slow
space-charge wave and the slow cyclotron wave. These are also
indicated in Figure 4.2. We see that there are large number of
interactions to be expected because of this coupling of propagating
waves.† Many of these interactions do not have counterparts in
the one-dimensional case considered in Chapter 3. Therefore,
the previous remarks concerning the importance of considering
the finite transverse boundaries are partly justified. The various
interactions indicated in Figure 4.2 will be considered in some
detail in Sections 4.1.1 and 4.1.2.

---

*The root given by Equation 4.10 will be termed the beam "cy-
clotron wave"; however, it should not be confused with the usual
physical picture of cyclotron and synchronous waves on a filamentary beam.[66] (See the discussion in Section 4.2.)

†For very low values of v_0, there is also a possibility of an interaction with the space-charge wave below ω_{ci}, which is not shown
in Figure 4.2.

The interactions resulting from a coupling of propagating waves are not the only possible ones, however. Since the beam is flowing through a medium that has a dielectric constant quite different from its free-space value, we should also expect that reactive-medium type of instabilities could arise in regions where the (effective) plasma dielectric constant is negative.

As a simple illustration, consider the case of zero magnetic field. For $B_0 = 0$, the dispersion equation becomes

$$p^2 = -\beta^2 \tag{4.12}$$

or

$$K_{\parallel}^0 = 0 \tag{4.13}$$

The solution given by Equation 4.12 represents the empty-waveguide modes below cutoff that in this case are independent of both the beam and the plasma. The dispersion equation (4.13) is identical to the one-dimensional dispersion equation analyzed in Section 3.1. Therefore, there is reactive-medium amplification for all frequencies below ω_{p0}, and this amplification becomes infinite at $\omega = \omega_{p0}$ (for a cold plasma[67]).

It will be shown in the following analysis that all amplification rates are finite for nonzero magnetic fields and for very weak beams ($\omega_{pb} \ll \omega_{ce}$). The example of zero magnetic field does indicate, however, that amplification rates are not necessarily finite for all beam-plasma systems of finite transverse dimensions (within a cold-plasma approximation). Another example where infinite amplification can occur is for very large magnetic fields, as considered in Section 4.3; in that case there is a critical value of beam density (beam perveance) for infinite amplification to be obtained. The question of how these results on "dense" beams extend to arbitrary values of the magnetic field is an important one that has not yet been fully worked out.

4.1.1 Space-Charge-Wave Interactions. Under the weak-beam assumption, the space-charge waves can be solved for by assuming $|\beta - \beta_e| \ll \beta_e$, as in the one-dimensional case. The dispersion equation (4.3) can be written in the form

$$(\beta - \beta_e)^2 = \frac{\beta^2 \beta_{pb}^2}{[\beta^2 - \beta_0^2(\omega)] K_{\parallel}} \tag{4.14}$$

where $\beta_0(\omega)$ is the dispersion of the plasma wave in the absence of the beam; that is,

$$\beta_0^2(\omega) = -p^2 \frac{K_\perp}{K_\parallel} \tag{4.15}$$

In Equation 4.14 we have placed $K_\perp^0 \simeq K_\perp$, a procedure consistent with the weak-beam assumption $(\omega_{pb} \ll \omega_{ce})$. Note that this is equivalent to the assumption that the ac motion of the beam electrons is along only the magnetic field, that is, that the electron beam is in confined flow.

Off-synchronism amplification. To solve for the beam waves from Equation 4.14, we can set $\beta = \beta_e$ on the right-hand side if we are far enough from the synchronous frequencies where $\beta_0(\omega) = \beta_e$ (see Equation 4.11). The frequencies of synchronism will be denoted as ω_1 and ω_2, corresponding to the circled regions 1 and 2 in Figure 4.2. We see that there are complex roots of β in the regions where

$$K_\parallel(\beta_e^2 - \beta_0^2) < 0 \qquad\qquad (4.16)$$

This can also be written in the form

$$\beta_e^2 K_\parallel + p^2 K_\perp < 0 \qquad\qquad (4.17)$$

Since $\beta = \beta_e \pm \delta(\omega)$ with $\delta(\omega) \to 0$ as $\omega_{pb} \to 0$, both roots of β must be in the lower-half β-plane for complex ω with sufficiently large negative imaginary part [greater than the order of Im (δv_0)]. Therefore, there are no absolute instabilities away from synchronism, and the roots with $\beta_i > 0$ are amplifying waves. The amplification mechanism is clearly a type of reactive-medium amplification, since it is necessary for either $K_\parallel$ or $K_\perp$, or both, to be negative (Inequality 4.17). From Inequality 4.16, we see that this reactive-medium amplification occurs in a low-frequency band from ω_{ci} to (approximately) ω_1 (see Figure 4.2), and in a high-frequency band from ω_{ce} to (approximately) ω_2. From Equation 4.14, it is clear that the maximum amplification rate must occur near to the synchronous frequencies ω_1 and ω_2, where $\beta_e = \beta_0(\omega)$.

Interaction at $\omega = \omega_2$. It has been noted by Kino and his associates[30] and Chodorow, Eidson, and Kino[33] that the dispersion equation (4.14) is identical to that of a backward-wave oscillator in the region where $K_\parallel > 0$, and where the "space-charge parameter"[47] QC of the equivalent backward-wave oscillator is equal to zero. The determinantal equation for a backward-wave oscillator is of the form[58]

$$(\beta - \beta_e)^2(\beta^2 - \beta_1^2) = 2\beta_e \beta_1 \beta^2 C^3 \qquad\qquad (4.18)$$

where $\beta_1(\omega)$ is the propagation constant of the circuit. For small C and for frequencies close to synchronism, Equation 4.18 can be written as

$$(\beta - \beta_e)^2(\beta - \beta_1) = \beta_e^3 C^3$$

Comparing Equations 4.18 and 4.14, we see that Pierce's C parameter for the beam-plasma system is given by

$$C^3 = \frac{\beta_{pb}^2}{2\beta_e\beta_0 K_{\parallel}} \simeq \frac{\beta_{pb}^2}{2\beta_e^2 K_{\parallel}} \qquad (4.19)$$

with $K_{\parallel}$ and β_e being evaluated at the intersection frequency ω_2.

The space-charge parameter QC in Pierce's theory is also a measure of the "coupling capacitor" between the beam and the circuit.[47] In the beam-plasma system, this "coupling capacitor" is essentially infinite, and for this reason QC = 0 in the present case, even though the beam space-charge is accounted for.

In Appendix H it is shown that there is an absolute instability near the intersection frequency ω_2, as we would expect from the previous analogy with the backward-wave oscillator. In fact, if we identify the quasi-static potential φ with the circuit voltage in Pierce's formulation, the relations between this voltage and the beam current and velocity modulations are exactly the same as for a backward-wave oscillator that uses an electron beam in confined flow. Therefore, the results that have been computed on the starting length for oscillation in a backward-wave oscillator can be applied directly to the beam-plasma system. For zero velocity and current modulation on the beam at z = 0, and zero "circuit voltage" (potential φ) at z = L,[*] the minimum length for oscillation L_s is given by[58]

$$\frac{\beta_e L_s}{2\pi} = \frac{0.314}{C} \qquad (4.20)$$

The frequency at which oscillations start is determined by[58]

$$\beta_0(\omega) - \beta_e = 1.52(\beta_e C) \qquad (4.21)$$

that is, the oscillations start at a frequency slightly below the intersection frequency ω_2.

The procedure for computing this starting length for any given set of physical parameters is as follows: first, the frequency ω_2 is determined by solving for the highest frequency for which $\beta_0(\omega) = \beta_e$. This frequency is then used to evaluate the C parameter in

---

[*]It is implied here that the "circuit" is matched at z = 0 so that there is no reflection of electromagnetic power, and means that the forward-traveling circuit wave $[\beta \simeq -\beta_0(\omega)]$ is not excited. If there is some reflection of electromagnetic power at z = 0, the starting length will in general be less than that given by Equation 4.20.

Equation 4.19, and the starting length is determined using Equation 4.20.

We note from Equation 4.19 that large C values at $\omega = \omega_2$ are obtained when $\omega_{po} \gg \omega_{ce}$, since $K_{\parallel}(\omega = \omega_2)$ is very small in this limit. For this case we find that

$$\frac{\omega_2^2}{\omega_{po}^2} - 1 \simeq \left(\frac{\omega_{ce}^2}{\omega_{po}^2}\right)\left(1 + \frac{\omega_{po}^2}{p^2 v_0^2}\right)^{-1} \tag{4.22}$$

and therefore the starting length is

$$\frac{L_s}{\lambda_{pb}} \simeq 0.4 \frac{\left(\frac{\omega_{pb}}{\omega_{ce}}\right)^{\frac{1}{3}} \left(\frac{\omega_{ce}}{\omega_{po}}\right)}{\left(1 + \frac{\omega_{po}^2}{p^2 v_0^2}\right)^{\frac{1}{3}}} \tag{4.23}$$

where $\lambda_{pb} = 2\pi/\beta_{pb}$ is the beam space-charge wavelength. In the opposite limit, when $\omega_{ce} \gg \omega_{po}$, we have $K_{\parallel}(\omega_2) \simeq 1$, and therefore

$$\frac{L_s}{\lambda_{pb}} \simeq 0.4 \left(\frac{\omega_{pb}}{\omega_{ce}}\right)^{\frac{1}{3}} \tag{4.24}$$

To illustrate the transition between these two extremes, the starting length is plotted in Figure 4.3 as a function $\omega_{pe}^2/\omega_{ce}^2$ for $pv_0/\omega_{ce} = 1$. Physically, this is essentially a plot of starting length versus plasma density for a fixed magnetic field and geometry.

Interaction at $\omega = \omega_1$. The low-frequency synchronism is always below ω_{po} where $K_{\parallel}$ is negative (Figure 4.2). Near the synchronous frequency ω_1, the dispersion equation can be written in the form

$$(\beta - \beta_e)^2[\beta - \beta_0(\omega)] = -\beta_e^3 C^3 \tag{4.25}$$

with

$$C^3 = \frac{\beta_{pb}^2}{2\beta_e^2(-K_{\parallel})} \tag{4.26}$$

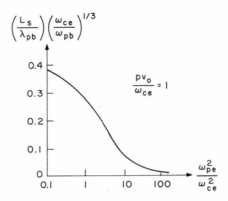

Figure 4.3. Starting length for oscillation at $\omega = \omega_2$.

and where $K_{\parallel}$ and β_e are evaluated at $\omega = \omega_1$. This dispersion equation is identical to that of the traveling-wave-tube amplifier with QC $\neq 0$.[47] In Appendix H, it is pointed out that there is no absolute instability, as would be expected because all three waves are forward-traveling waves ($\partial\omega/\partial\beta > 0$) when "uncoupled," that is, when C = 0. The maximum amplification occurs at the frequency $\omega = \omega_1$ where $\beta_0 = \beta_e$. This maximum amplification is given by

$$(\beta_i)_{max} = \frac{\sqrt{3}}{2} \beta_e C \tag{4.27}$$

As before, the value of this maximum amplification is determined by solving for ω_1 and using this frequency to determine C from Equation 4.26. The frequency ω_1 can be in the very low frequency regime where the ions are of importance (on the order of ω_0), or it can be just slightly less than (the smaller of) ω_{ce} and ω_{p0} for low values of p^2. As an illustration, the frequency ω_1 is plotted as a function of magnetic field $\omega_{ce}^2/\omega_{p0}^2$ for fixed density and geometry $(pv_0/\omega_{p0})^2 = 0.1$ in Figure 4.4. The value of the maximum amplification rate is also plotted in the figure as a function of $\omega_{ce}^2/\omega_{p0}^2$. We note that the amplification becomes very small when the parameters are such that the frequency of synchronism ω_1 is in the very low frequency regime. Mathematically, this is to be expected because C^3 is inversely proportional to $K_{\parallel}$, which becomes very large at low frequencies. Physically, this reduction in amplification at low frequencies comes about because of the "shorting out" of the low-frequency electric field by the "cold" plasma electrons.

4.1.2 Cyclotron-Wave Interactions. For a weak beam the slow cyclotron wave can be solved for by assuming that $\beta - \beta_{sc} = \delta$, with $\delta \ll \beta_{sc}$, where

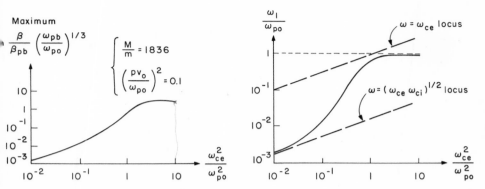

Figure 4.4. Amplification at $\omega = \omega_1$.

$$\beta_{sc} = \beta_e + \beta_{ce} \tag{4.28}$$

The dispersion equation (4.3) can then be written in the form

$$\frac{\dfrac{\beta_{pb}^2}{2\beta_{ce}}}{\beta - \beta_{sc}} = \frac{K_{\parallel}}{p^2}[\beta^2 - \beta_0^2(\omega)] \tag{4.29}$$

where $\beta_0(\omega)$ is the unperturbed plasma wave (Equation 4.15), and we have placed $K_{\parallel}^0 \simeq K_{\parallel}$ since we are assuming $\omega_{pb}^2 \ll \omega_{ce}^2$. Note that in this case we are effectively ignoring the <u>longitudinal</u> ac motion of the beam electrons.

We see from Equation 4.29 that β is always real far from the synchronous frequencies where $\beta_{sc} = \beta_0$, and therefore there are no amplifying waves or instabilities away from synchronism. For frequencies close to synchronism, the dispersion equation can be written as

$$(\beta - \beta_{sc})(\beta - \beta_0) = \frac{p^2\beta_{pb}^2}{4\beta_{ce}\beta_0 K_{\parallel}}$$

$$\overset{\Delta}{=} c_0^2 \tag{4.30}$$

where the sign of c_0^2 depends on the sign of $K_{\parallel}$. Once again, we can evaluate c_0^2 at the <u>synchronous</u> frequency for a weak beam. The three synchronous frequencies will be denoted as ω_3, ω_4, and ω_5 corresponding to the regions labeled 3, 4, and 5 in Figure 4.2, and the corresponding values of c_0 will be labeled as c_{03}, c_{04}, and c_{05}.

The dispersion equation (4.30) is of exactly the form that is obtained from weak-coupling theory (see Section 2.7.1). Therefore, we can immediately conclude that the interactions at ω_3 and ω_4 correspond to amplifying waves because $c_0^2 < 0$ and $\partial\beta_0/\partial\omega > 0$; the interaction at ω_5 corresponds to an absolute instability since $\partial\beta_0/\partial\omega < 0$ and $c_0 > 0$.* All of these results are exactly what one would have expected at the outset from the theory of backward-wave oscillators and traveling-wave-tube amplifiers.

It is interesting to note that c_0^2 is proportional to p^2; that is, the interactions tend to be stronger for the higher-order modes. The opposite is usually the case for the space-charge wave interactions. There the interactions were usually stronger for the lower-order modes that have a smoother variation of the longitudinal electric field over the cross section (see, for example, Equation 4.23). However, it is perhaps reasonable that the transverse interactions in a filled guide should be stronger for the higher-order modes, since the transverse electric field is the order of p times the potential φ. It can be shown that the electromagnetic power in a propagating plasma wave is proportional to an integral of $|\varphi|^2$ over the cross section;[61,65] therefore, the higher-order plasma modes have a larger maximum amplitude of the transverse electric field per unit power and should therefore couple more strongly to the beam electron's cyclotron motion.[66]

To find the amplification rates and starting lengths for these interactions, one must first solve for the three synchronous frequencies from $\beta_e + \beta_{ce} = \beta_0(\omega)$ and then use this frequency to evaluate c_0^2 in Equation 4.30.

Interaction at $\omega = \omega_3$. The interaction near the ion cyclotron frequency (see Figure 4.2) occurs at a frequency

$$\omega_3 = \frac{\omega_{ci}}{\left[1 + \left(\frac{m}{M}\right)\left(\frac{p^2}{\beta_{ce}^2}\right)\right]^{\frac{1}{2}}} \qquad (4.31)$$

for an assumption that $\omega_{ci} \ll \omega_{pi}$. As mentioned before, this coupling leads to amplifying waves; the maximum amplification rate occurs at $\omega = \omega_3$ and is equal to

$$(\beta_i)_{max} = c_{03} = \frac{1}{2}\beta_{pb}\left(\frac{\omega_{ci}}{\omega_{po}}\right)\frac{\frac{p}{\beta_{ce}}}{\left[1 + \left(\frac{m}{M}\right)\left(\frac{p^2}{\beta_{ce}^2}\right)\right]^{\frac{1}{2}}} \qquad (4.32)$$

---

*If a similar analysis is applied to the fast cyclotron wave ($\beta = \beta_e - \beta_{ce}$), it is easily shown that all synchronisms lead to stable waves, as expected.

This growth rate is usually extremely small. Even for the very high order modes, with $p^2/\beta_{ce}^2 \gg M/m$, this growth rate is on the order of 3×10^{-3} cm^{-1} for protons with $n_b/n_p = 10^{-2}$ and a magnetic field of 1000 gauss.

Interaction at $\omega = \omega_4$. The amplification that results from the coupling at $\omega = \omega_4$ can occur at frequencies between ω_0 and the lower of ω_{po} and ω_{ce} (Figure 4.2). The frequency ω_4 tends to ω_0 as p^2 becomes very large $[p^2 \gg \beta_{ce}^2(\omega_0^2/\omega_{po}^2)]$. The maximum amplification rate in this limit becomes

$$(\beta_i)_{max} = c_{04} = \frac{1}{2}p\left(\frac{\omega_{pb}}{\omega_{po}}\right)\left(\frac{\omega_0}{\omega_{ce}}\right) \tag{4.33}$$

which now increases indefinitely as $p \to \infty$ (although care must be exercised since our weak-beam assumption breaks down when c_{04} is on the order of β_{ce} or larger).

An analytical solution can also be obtained when $\omega_{ce} \gg \omega_{po}$. In this case we find

$$\omega_4 \simeq \omega_{po} \frac{\left[1 + \left(\frac{m}{M}\right)\left(\frac{p^2}{\beta_{ce}^2}\right)\right]^{\frac{1}{2}}}{\left(1 + \frac{p^2}{\beta_{ce}^2}\right)^{\frac{1}{2}}} \tag{4.34}$$

and that

$$(\beta_i)_{max} = \frac{1}{2}\beta_{pb}\left[1 + \left(\frac{m}{M}\right)\left(\frac{p^2}{\beta_{ce}^2}\right)\right]^{\frac{1}{2}} \tag{4.35}$$

This passes over into Equation 4.33 for $p^2 \gg (M/m)\beta_{ce}^2$ because $\omega_0 \to \omega_{pi}$ for $\omega_{ce} \gg \omega_{po}$. To find the amplification rate for arbitrary values of ω_{po} and ω_{ce}, we must, in general, solve a fourth-order equation to determine ω_4.

Interaction at $\omega = \omega_5$. The absolute instability that results from a coupling of the backward plasma wave with the slow cyclotron wave occurs at a frequency ω_5 between the higher of ω_{po} and ω_{ce} and the frequency ω_+ (Figure 4.2). For $\omega_{ce} \gg \omega_{po}$, the frequency is on the order of ω_{ce}, and therefore the perturbation in the wave number ($\delta = c_{05}$) at ω_5 is[36]

$$\delta^2 = p^2\left(\frac{\omega_{pb}^2}{8\omega_{ce}^2}\right) \tag{4.36}$$

The weak-coupling description of this interaction is clearly valid for a weak beam, and therefore the starting length for oscillation when we neglect any reflections is[2,36,59]

$$L_s = \frac{1}{4}\left(\frac{2\pi}{\delta}\right) = \frac{\pi}{2\delta} \tag{4.37}$$

Morse[36] has considered this interaction in some detail, and he has pointed out that even for the dominant mode (lowest p) this starting length can be on the order of only a few cyclotron wavelengths. Once again, the strongest interactions are for the higher-order modes, and these have the shortest starting lengths.

For $\omega_{ce} \ll \omega_{p0}$, the interaction occurs at the frequency

$$\omega_4 = \omega_{p0}\left(1 + \frac{\dfrac{\omega_{ce}^2}{\omega_{p0}^2}}{1 + \dfrac{\omega_{p0}^2}{p^2 v_0^2}}\right)^{\frac{1}{2}} \tag{4.38}$$

and therefore the perturbation in the wave number is now given by

$$\delta^2 = \frac{1}{4}p^2\left(\frac{\omega_{pb}^2 \omega_{p0}}{\omega_{ce}^3}\right)\left(1 + \frac{\omega_{p0}^2}{p^2 v_0^2}\right) \tag{4.39}$$

In closing, it might be appropriate to add a few additional comments about these cyclotron-wave interactions. The simple theory given here indicates that these interactions are strongest for the higher-order modes that have a large number of variations over the transverse cross section. However, in this simplified model it was assumed that the unperturbed motion of the beam electrons is that of perfect rectilinear flow, and that the plasma electrons do not have any transverse temperature, that is, they also do not execute any unperturbed motion about the magnetic field. Intuitively, it would seem that these two factors, as well as the inevitable inhomogeneities in the beam and plasma densities over the cross section, would tend to put an upper bound on the radial wave number p for which a strong coherent interaction could be obtained. Moreover, it might also be conjectured that when the coherent transverse motion is of a large enough amplitude so that the radius of gyration of the electrons is on the order $1/p$, the interaction would rapidly die out. In brief, there is reason to suspect that the mathematical idealizations used here are particularly poor for these higher-order modes, and therefore the practical importance of these interactions with the higher-order modes is not entirely clear.

In all of the interactions described in this section, it was generally true that the high-frequency instabilities (at frequencies
characteristic of the plasma electrons) were much stronger than
any of the low-frequency instabilities (at frequencies characteristic of the plasma ions). This general trend was particularly pro-
nounced in the case of the longitudinal space-charge wave instabilities. Once again, the physical reason for this behavior is that
cold electrons tend to "short out" any low-frequency electric fields,
and particularly those fields that are aligned with the steady magnetic field. In the one-dimensional case considered in Chapter 3,
we found that a strong interaction with the ions could be obtained
only when the plasma electrons are relatively hot. The extension
of these warm-plasma results to the case of a finite geometry is
given in Chapter 5.

4.2 Thin-Beam Interactions

In this section we shall consider the interaction of a very thin
"pencil" electron beam with a homogeneous plasma that fills a
circular waveguide (Figure 4.5). Both the circularly symmetric
modes ($n = 0$) and the modes of one azimuthal variation ($n = \pm 1$)
will be investigated.

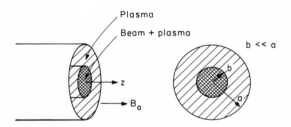

Figure 4.5. Thin-beam system.

The quasi-static dispersion equation for the system shown in
Figure 4.5 is derived in Appendix G. For an assumed angular
variation of the form $e^{jn\phi}$, it is

$$
K_\perp^0 pb \left[\frac{J'_n(pb)}{J_n(pb)} \right] + n \left[\frac{\omega_{pb}^2 \omega_{ce}}{\omega_d(\omega_d^2 - \omega_{ce}^2)} \right] = K_\perp qb \left\{ \frac{N'_n(qb) - \left[\frac{N_n(qa)}{J_n(qa)} \right] J'_n(qb)}{N_n(qb) - \left[\frac{N_n(qa)}{J_n(qa)} \right] J_n(qb)} \right\}
$$

$$(4.40)$$

where p is the transverse wave number inside the beam region
(Equation 4.3), $\omega_d = \omega - \beta v_0$, and

$$q^2 = -\beta^2 \frac{K_{\parallel}}{K_{\perp}} \qquad (4.41)$$

is the transverse wave number in the plasma (all other symbols are defined in Section 4.1). When q^2 is approximately real and negative, a more convenient form of this dispersion equation is

$$K_{\perp}^0 pb \left[\frac{J_n'(pb)}{J_n(pb)}\right] + \left[n \frac{\omega_{pb}^2 \omega_{ce}}{\omega_d(\omega_d^2 - \omega_{ce}^2)}\right] = K_{\perp} q'b \left\{ \frac{K_n'(q'b) - \left[\frac{K_n(q'a)}{I_n(q'a)}\right] I_n'(q'b)}{K_n(q'b) - \left[\frac{K_n(q'a)}{I_n(q'a)}\right] I_n(q'b)} \right\}$$

$$(4.42)$$

where K_n and I_n are the modified Bessel functions,[54] and $(q')^2 = -q^2$.

A general solution for $\beta(\omega)$ or $\omega(\beta)$ from the above is clearly quite difficult. Allen and Kino and coworkers[30-32] have performed some computations on the space-charge-wave interaction of a weak beam near the synchronism with the backward plasma wave (analogous to the interaction labeled 2 in Figure 4.2). These computations showed that there are many peaks of β_i vs. real ω in the frequency range between ω_+ and (the higher of) ω_{po} or ω_{ce} when the beam does not fill the waveguide. These peaks can be thought of as arising from the coupling of a single "space-charge wave mode" (that is, the beam waves in the absence of the plasma) with many propagating "plasma modes" (that is, the plasma waves in the absence of the beam). This multitude of peaks did not arise in the case of the filled waveguide considered in Section 4.1, because the transverse variation of the beam modes, which is then just $J_n[\epsilon_{nm}(r/a)]$, "fits" only the single plasma mode that has the same transverse variation. In a sense, we could say that the beam modes are orthogonal to all except one of the plasma modes in the case for a filled guide; this orthogonality would be quite apparent in a variational formulation of the interaction.[49] There would be no such orthogonality in the unfilled case, as is demonstrated by Allen and Kino's computations. This lack of orthogonality should also be obtained for the other coupling regions (analogous to those indicated in Figure 4.2) for a beam that does not fill the guide, although there have been few computations at the present time.*

In the following only the lower-order modes that have relatively constant fields over the cross section of a very thin beam will be considered (pb ≪ 1). It will be shown that only the space-charge

---

*It should be mentioned that Maxum[68] has performed some computations on the double-stream interaction between two beams with different cross sections.

wave interactions, and not the cyclotron interactions, are obtained from the circularly symmetric (n = 0) mode with pb $\ll$ 1. This is to be expected because the electric field is purely longitudinal at the position of the beam (r = 0) for this mode. These space-charge-wave interactions are qualitatively similar to the case of a filled waveguide (except for the multitude of synchronisms), and a "reduction factor" is derived that enables the results of Section 4.1 to be applied to the present case. The n = ±1 modes with pb $\ll$ 1 correspond (in the absence of the plasma) to the usual cyclotron and synchronous modes of a filamentary beam. In the presence of a plasma these modes can be unstable because of a transverse-reactive-medium effect as well as the usual synchronisms with propagating plasma waves.

4.2.1 n = 0 Modes. The dispersion equation (4.40) for the n = 0 modes becomes, when we assume qb $\ll$ 1 and pb $\ll$ 1,

$$\frac{1}{2}(pb)^2 K_\perp^0 \simeq K_\perp qb \left\{ \frac{\dfrac{2}{\pi qb}\left[\dfrac{J_0(qa)}{N_0(qa)}\right] - \dfrac{1}{2}qb}{\dfrac{2}{\pi}\ln\left(\dfrac{1}{qb}\right)\left[\dfrac{J_0(qa)}{N_0(qa)}\right] + 1} \right\} \tag{4.43}$$

If we neglect further terms on the order of $(pb)^2$ and $(qb)^2$ and assume that $K_\perp \neq 0$, Equation 4.43 becomes

$$\frac{1}{2}\frac{(\beta b)^2 \beta_{pb}^2}{(\beta - \beta_e)^2} \simeq K_\perp \left\{ \frac{\dfrac{2}{\pi}\left[\dfrac{J_0(qa)}{N_0(qa)}\right]}{\dfrac{2}{\pi}\ln\left(\dfrac{1}{qb}\right)\left[\dfrac{J_0(qa)}{N_0(qa)}\right] + 1} \right\} \tag{4.44}$$

In deriving this equation, we have used the following expansions for the Bessel functions valid for small values of the argument:[54]

$$J_0(x) \simeq 1 - \frac{1}{4}x^2 \tag{4.45}$$

$$N_0(x) \simeq \frac{2}{\pi}\ln x \tag{4.46}$$

Reactive-medium amplification. Except for the frequencies close to synchronism with a plasma wave, where $J_0(qa) \simeq 0$, from Equation 4.44 the solution for the space-charge waves is

$$(\beta - \beta_e)^2 = \frac{\frac{1}{2}\beta_{pb}^2 b^2 \beta_e^2}{K_\perp}\ln\left(\frac{1}{q_e b}\right) \tag{4.47}$$

where q_e is the value of q obtained when $\beta = \beta_e$ in Equation 4.41. When q_e^2 is negative, which occurs when the plasma is nonpropagating, a similar expansion of Equation 4.42 gives

$$(\beta - \beta_e)^2 = \frac{\frac{1}{2}\beta_{pb}^2 b^2 \beta_e^2}{K_\perp} \ln\left(\frac{1}{q_e^! b}\right) \tag{4.48}$$

with $(q_e^!)^2 = -q_e^2$. From the Equations 4.47 and 4.48 we see that reactive-medium amplification is obtained in the frequency regions where $K_\perp$ is negative. In terms of the frequencies defined in Figure 4.2, there is reactive-medium amplification in a low-frequency band ω_{ci} to ω_0 and in a high-frequency band ω_{ce} to ω_+. Note that this frequency band is slightly different from the case for a reactive-medium band of a particular mode in the filled guide.

The result given by Equations 4.47 and 4.48 is not valid when $K_{||} = 0$ because $q \to 0$ and the $J_0(qa)/N_0(qa)$ term must be retained in Equation 4.44. In fact, at $\omega = \omega_{p0}$ we have

$$(\beta - \beta_e)^2 = -\frac{1}{2}\beta_{pb}^2 b^2 \beta_e^2 \left(\frac{\omega_{p0}^2 + \omega_{ce}^2}{\omega_{ce}^2}\right) \ln\left(\frac{a}{b}\right) \tag{4.49}$$

These results are also invalid near $K_\perp = 0$ because $q \to \infty$ at these frequencies for a finite β. From Equation 4.40 we find that at $K_\perp = 0$ the lowest-order mode has $(pb)^2 K_\perp^0 = 0$, or

$$(\beta - \beta_e)^2 = \frac{\beta_{pb}^2}{(K_{||})_0} \tag{4.50}$$

where $(K_{||})_0$ is evaluated at the frequencies where $K_\perp = 0$; that is, $\omega = \omega_0$ and $\omega = \omega_+$.

Synchronous frequencies. We now consider the synchronous frequencies where $J_0(q_e a) \simeq 0$. We define $\beta_0(\omega)$ as the dispersion of one of the propagating plasma modes in the absence of the electron beam; that is,

$$\beta_0^2 = -q_n^2 \left(\frac{K_\perp}{K_{||}}\right) \tag{4.51}$$

where $q_n a$ is one of the roots of $J_0(q_n a) = 0$. For ω and β in the neighborhood of such a propagating wave, we have

$$\frac{J_0(qa)}{N_0(qa)} \simeq \left[\frac{aJ_0^!(q_n a)}{N_0(q_n a)}\right]\left(\frac{\partial q}{\partial \beta}\right)_0 [\beta - \beta_0(\omega)] \simeq -\frac{2}{\pi\beta_0 N_0^2(q_n a)}(\beta - \beta_0) \tag{4.52}$$

The latter form is derived by using the Wronskian relation,[54]

$$J_0(x)N_0'(x) - J_0'(x)N_0(x) = \frac{2}{\pi x} \qquad (4.53)$$

By using Equation 4.52 in Equation 4.44, we can write a dispersion equation that is valid near the synchronous frequencies where $\beta_0 \simeq \beta_e$:

$$(\beta - \beta_e)^2(\beta - \beta_0) = \left(\frac{\beta_{pb}^2 \beta_e}{2K_{\parallel}}\right)\left[\frac{\pi^2}{4} q_n^2 b^2 N_0^2(q_n a)\right] \qquad (4.54)$$

Comparing Equation 4.54 with the dispersion equation for space-charge-wave interactions in a filled waveguide, Equation 4.14, we see that the two dispersion equations are of exactly the same form near the synchronous frequencies, the only difference being the positive factor less than unity in the brackets in Equation 4.54. Therefore, these interactions are of exactly the same type as in Section 4.1.1, and, in fact, the results given there can be adapted to the present case by introducing the "reduction factor" defined by

$$r_n^3 = \left(\frac{\pi^2}{4}\right)(q_n b)^2 N_0^2(q_n a) \qquad (4.55)$$

Note that there is a different reduction factor for each radial plasma mode. The C parameter in the present case is clearly just r_n times the C parameter determined in Section 4.1.1.* Since $q_n b \ll 1$ for the lower-order plasma modes, this reduction factor is less than unity for these modes. (Note, however, that we have <u>assumed</u> $qb \ll 1$ at the outset, and therefore the higher-order modes, where $qb \gtrsim 1$, cannot be considered in the present formulation.)

As an example, consider the starting length for backward-wave oscillations at $\omega = \omega_2$ when $\omega_{ce} \ll \omega_{p0}$. From Equations 4.23 and 4.55 this starting length is

$$\frac{L_s}{\lambda_{pb}} = 0.3 \frac{\left(\dfrac{\omega_{pb}}{\omega_{ce}}\right)^{\frac{1}{3}}\left(\dfrac{\omega_{ce}}{\omega_{p0}}\right)}{[N_0(q_n a)]^{\frac{2}{3}} (q_n b)^{\frac{2}{3}}\left(1 + \dfrac{\omega_{p0}^2}{q_n^2 v_0^2}\right)} \qquad (4.56)$$

for the nth plasma mode. If we write Equation 4.56 in an unnormalized form, we see that the starting length is inversely proportional to $(\beta_{pb}b)^{2/3}$. The quantity $(\beta_{pb}b)^2$ is proportional to the beam perveance K (see Equation 4.84), and therefore this starting length

---

*Note that wherever p appears in Section 4.1, it should be replaced by q_n.

is inversely proportional to $K^{1/3}$. This indicates that for a very thin electron beam it is the beam perveance $b^2 n_b$ that is of primary importance in this interaction rather than only the beam density n_b.

It is interesting to note that the circularly symmetric interactions with pb $\ll$ 1 are only of the "space-charge" type; that is, we did not find any interactions caused by synchronisms with a "slow cyclotron wave" with $\beta \simeq \beta_e + \beta_{ce}$. This is not too surprising, since the circularly symmetric plasma modes have only a longitudinal component of the electric field at $r = 0$ and therefore should not couple with the transverse motion of a thin beam. In the case of the filled waveguide, however, even the circularly symmetric modes have an interaction near $\beta_e + \beta_{ce}$. If we consider the beam waves in the absence of the plasma in the unfilled waveguide, a closer examination shows that all of these "cyclotron" modes for $n = 0$ are "bodylike" waves in the sense that pb is always greater than the first zero of $J_0(x)$. The validity of this statement follows directly from the well-known analysis of the $n = 0$ modes in an unfilled plasma waveguide.[1,29] The beam waves can be obtained from the plasma solutions by a simple transformation of coordinates, and one finds that the backward plasma wave (a body wave) goes into the $n = 0$ "cyclotron" beam wave. Therefore, these cyclotron interactions in the case for $n = 0$ must have a very nonuniform field over the cross section of the beam. The "true" cyclotron wave interactions in the case of the thin beam are of the $n = 1$ type, as will be shown in the next section.

4.2.2 n = ±1 Modes. The dispersion equation for the $n = +1$ mode becomes, when we assume pb $\ll$ 1 and qb $\ll$ 1,

$$K_\perp^0 + \frac{\omega_{pb}^2 \omega_{ce}}{\omega_d(\omega_d^2 - \omega_{ce}^2)} = -K_\perp \left\{ \frac{1 + \frac{q^2 b^2}{2} \ln qb - \frac{\pi}{4}(q^2 b^2)\left[\frac{N_1(qa)}{J_1(qa)}\right]}{1 - \frac{q^2 b^2}{2} \ln qb + \frac{\pi}{4}(q^2 b^2)\left[\frac{N_1(qa)}{J_1(qa)}\right]} \right\} \quad (4.57)$$

where we have made use of the following expansions of the Bessel functions in the limit of small argument:[54]

$$J_1(x) \simeq \frac{1}{2}x \quad (4.58)$$

$$N_1(x) \simeq -\frac{2}{\pi x} + \frac{x}{\pi} \ln x \quad (4.59)$$

This dispersion equation can be written in the form

$$\frac{\omega_{pb}^2}{\omega_d(\omega_d + \omega_{ce})} = \frac{2K_\perp}{1 + \frac{\pi}{4}(q^2 b^2)\left[\frac{N_1(qa)}{J_1(qa)}\right] + \frac{1}{2}q^2 b^2 \ln\left(\frac{1}{qb}\right)} \quad (4.60)$$

The dispersion of the $n = -1$ mode follows from Equation 4.60 by letting $B_0 \to -B_0$.

Transverse reactive-medium amplification. As long as we are not too close to synchronism with a plasma wave, where $J_1(qa) \simeq 0$, Equation 4.60 takes the simple form first derived by Kino and Gerchberg:[38]

$$(\beta - \beta_e)(\beta - \beta_e - \beta_{ce}) = \frac{\beta_{pb}^2}{2K_\perp} \tag{4.61}$$

and therefore

$$\beta = \beta_e + \frac{\beta_{ce}}{2} \pm \left(\frac{\beta_{ce}^2}{4} + \frac{\beta_{pb}^2}{2K_\perp}\right)^{\frac{1}{2}} \tag{4.62}$$

This dispersion equation can be derived in a very simple fashion by noting that the assumptions $qb \ll 1$ and $pb \ll 1$ imply that the potentials in the region of the beam are approximately solutions of the two-dimensional Laplace equation, $\nabla_T^2 \phi \simeq 0$. For the $n = 1$ modes, the appropriate solution inside the beam is $(r/b) \exp j\phi$, while outside it is $(b/r) \exp j\phi$. The application of the boundary conditions, following the method outlined in Appendix G, immediately yields the dispersion equation (4.61).[38]

The beam waves in the absence of the plasma ($K_\perp = 1$) are sketched in Figure 4.6. In the limit of a very weak beam ($\omega_{pb} \ll \omega_{ce}$), these waves pass over into the well-known cyclotron and synchronous waves of a filamentary beam.[66] The active waves (those that carry negative small-signal energy) are the slow cyclotron wave with $\beta \simeq \beta_e + \beta_{ce}$, which has a left-polarized sense of rotation in time ($n = +1$), and the slow synchronous wave with $\beta \gtrsim \beta_e$, which rotates in the

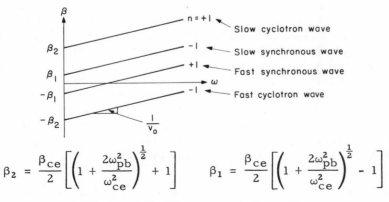

$$\beta_2 = \frac{\beta_{ce}}{2}\left[\left(1 + \frac{2\omega_{pb}^2}{\omega_{ce}^2}\right)^{\frac{1}{2}} + 1\right] \qquad \beta_1 = \frac{\beta_{ce}}{2}\left[\left(1 + \frac{2\omega_{pb}^2}{\omega_{ce}^2}\right)^{\frac{1}{2}} - 1\right]$$

Figure 4.6. Beam waves ($n = 1$).

right-polarized sense in time (n = -1).* For cases in which ω_{pb} is comparable with ω_{ce}, the results given here provide an extension of the filamentary-beam results.

In the presence of the plasma, we see that complex roots of β for real ω are obtained for both n = +1 and n = -1 when

$$- \frac{2\omega_{pb}^2}{\omega_{ce}^2} < K_\perp < 0 \tag{4.63}$$

The complex root with $\beta_i > 0$ is an amplifying wave, since for $\omega_i \to -\infty$ both roots of β from Equation 4.61 are in the lower-half β-plane. By the same argument, there can be no absolute instabilities (away from synchronisms, at least) because both roots of β are "downstream" waves. The amplification rate, according to this simple formulation, becomes <u>infinite</u> at the frequencies where $K_\perp \to 0^-$, which occurs at $\omega_0 = [(1/\omega_{ce}\omega_{ci}) + 1/\omega_{pi}^2]^{-1/2}$ and $\omega_+ = (\omega_{p0}^2 + \omega_{ce}^2)^{1/2}$. The amplification mechanism is clearly of the reactive-medium type because $K_\perp$ must be negative (Equation 4.63). The amplification extends over a narrow low-frequency band below ω_0, and over a high-frequency band up to ω_+ (Figure 4.7).

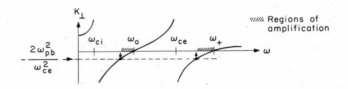

Figure 4.7. Regions of transverse reactive-medium amplification.

This transverse reactive-medium amplification is obtained even when there is no steady magnetic field (B_0 = 0).[38] In this case there is amplification over the entire range where $K_\parallel$ is negative ($\omega < \omega_{p0}$). The plasma can now be considered as an isotropic dielectric medium with $\overline{D} = \epsilon_0 K_\parallel \overline{E}$. For a very thin beam, the electric field inside the beam is a uniform transverse field, while just outside it has a dipole structure (Figure 4.8). If the beam is displaced in the transverse direction as a whole, an effective surface charge is created that results in the displacement vector $\overline{D}$ being as shown. When $K_\parallel$ is negative, $\overline{E}$ is antiparallel to $\overline{D}$, and it therefore exerts a force on the beam electrons that tends to drive the beam <u>farther</u> away from its equilibrium position. The result is that this "hoselike" perturbation

---

*The active waves are those which have $\omega/\beta < v_0$, as discussed in Appendix F.

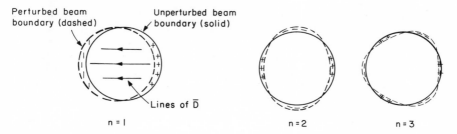

Figure 4.8. Beam boundary perturbations for transverse instability ($B_0 = 0$).

of the beam is unstable. The physical description of the instability is more complicated when there is a magnetic field present; however, the mechanism is essentially that just described.

It should be mentioned that Equation 4.61 is also obtained <u>for all modes with n > 1</u> when the thin-beam approximation is made, as can be verified by expanding Equation 4.40 for arbitrary n > 0. The perturbation in the beam shape corresponding to these higher-order angular modes is illustrated in Figure 4.8 for n = 2 and n = 3. The physical description of these instabilities is much the same as for n = 1; however, note that these instabilities for n > 1 tend to form the beam into "lobes" or "spokes" and do not result in a bodily motion of the beam center of mass away from the axis.

The previous result of infinite amplification at $K_\perp = 0$ for the n = ±1 modes violates the thin-beam assumption, since we assumed $|\beta b| \ll 1$ at the outset. In fact, even for finite β we see from Equation 4.41 that q → ∞ as $K_\perp$ → 0, and therefore the assumption of qb ≪ 1 is particularly poor in this limit. The preceding analysis does show, however, that the maximum amplification rate must <u>tend</u> to infinity as we let the beam radius b approach zero. We can evaluate the amplification rate at $K_\perp = 0$ if we assume that b is extremely small in order that $\lceil \beta \rfloor$ be much larger than β_{ce} or β_{pb} (high-gain limit). The dispersion equation (4.40) at $K_\perp = 0$ becomes

$$pb\left[\frac{J_1'(pb)}{J_1(pb)}\right] = \pm \frac{\beta_{ce}}{\beta - \beta_e} \tag{4.64}$$

Therefore, for $|\beta| \gg \beta_{ce}$, we must have pb ≃ 1.84 for the lowest-order mode. From Equation 4.3, we find that

$$\beta^4 \simeq \frac{p^2 \beta_{pb}^2}{(K_\parallel)_0} \tag{4.65}$$

where $(K_\parallel)_0$ is the value of $K_\parallel$ at the two frequencies where $K_\perp = 0$. As a numerical example, consider an electron-proton plasma with

$n_p = 10^{12}/cc$, $B_0 = 1000$ gauss, $n_b = 10^{10}/cc$, $v_0/c = 0.2$, and $b = 1$ mm. For these numbers the high-frequency amplification at ω_+ is on the order of 5 cm^{-1}, and the low-frequency amplification at ω_0 is on the order of 0.25 cm^{-1}. Actually, for this beam radius the amplification rates computed from Equation 4.65 are too low for the "high-gain" theory to apply, but they do illustrate the general rule that the amplification is appreciably reduced at low frequencies because of the "shorting out" effect of the cold plasma electrons.

Synchronous frequencies. As in the n = 0 case, a more careful expansion of the dispersion equation (4.60) must be made close to the synchronisms with propagating plasma waves, where $J_1(qa) \simeq 0$. Note, however, that if β is appreciably complex in some frequency range because of the reactive-medium effect, then there is no alteration in the previous results, since q is complex in Equation 4.60, and therefore the $N_1(qa)/J_1(qa)$ term can still be ignored (for a very thin beam) since it never tends to infinity. In the following, we shall consider only the case of a very weak beam ($\omega_{pb} \ll \omega_{ce}$) for which the reactive-medium amplification extends over only a very small range of frequency (Figure 4.7). Near a propagating plasma wave, for which

$$\beta_0^2(\omega) = -q_n^2 \frac{K_\perp}{K_\parallel} \tag{4.66}$$

and $q_n a$ is one of the roots of $J_1(q_n a) = 0$, we have

$$J_1(qa) \simeq a J_1'(q_n a)\left(\frac{\partial q}{\partial \beta}\right)_0 (\beta - \beta_0)$$

$$\simeq -\frac{2}{\pi \beta_0 N_1(q_n a)} (\beta - \beta_0) \tag{4.67}$$

The latter form follows from the Wronskian relation[54]

$$J_1(x)N_1'(x) - J_1'(x)N_1(x) = \frac{2}{\pi x} \tag{4.68}$$

From Equation 4.60, we find that the dispersion equation near such a synchronism can be written in the form

$$\frac{\beta_{pb}^2}{(\beta - \beta_e)(\beta - \beta_e - \beta_{ce})} = -\frac{16}{\pi^2}\left[\frac{K_\perp}{(q_n b)^2 N_1^2(q_n a)\beta_0}\right](\beta - \beta_0) \tag{4.69}$$

for n = +1 and the nth plasma mode.

The interaction with the slow cyclotron wave (Figure 4.6), for such a weak beam, is described by

$$(\beta - \beta_e - \beta_{ce})(\beta - \beta_0) = \left(\frac{\beta_{pb}^2 q_n^2}{4\beta_{ce}\beta_0 K_{\parallel}}\right)\left[\frac{\pi^2}{4} q_n^2 b^2 N_1^2(q_n a)\right] \qquad (4.70)$$

By comparing Equation 4.70 with the filled-guide dispersion equation near the cyclotron instabilities, Equation 4.30, we see that they are of exactly the same form except for the "reduction factor":

$$r_n^2 = \frac{\pi^2}{4} q_n^2 b^2 N_1^2(q_n a) \qquad (4.71)$$

in the "coupling coefficient" c_0^2, defined in Equation 4.30. As in the case for $n = 0$, we can carry over all of the results from Section 4.1 to the present case by multiplying the "coupling coefficient" c_0 by the reduction factor r_n (and noting that $p \to q_n$ in the present case). The very thin filamentary beam with weak space charge, therefore, exhibits traveling-wave amplification at the frequencies ω_3 and ω_4 indicated in Figure 4.2, and backward-wave oscillation at the frequency ω_5. Note that there are three interactions for every propagating plasma mode q_n.

It can be shown from Equation 4.69 that the interactions with the fast synchronous wave ($\beta \simeq \beta_e$ for $n = +1$) do not produce any instabilities, as would be expected because it is a positive-energy (passive) wave. The resulting dispersion equation near any synchronism is of exactly the same form as those discussed in Section 2.7.1. The same is true of the fast cyclotron wave. The interaction with the slow synchronous wave, however, should produce instabilities since it is a negative-energy (active) wave. The dispersion equation close to a synchronism with the slow synchronous wave, where $\beta_e \simeq \beta_0$ and $n = -1$, is of the form

$$(\beta - \beta_e)(\beta - \beta_0) = \left(\frac{\beta_{pb}^2 q_n^2}{4\beta_{ce}\beta_0 K_{\parallel}}\right)\left[\frac{\pi^2}{4} q_n^2 b^2 N_1^2(q_n a)\right] \qquad (4.72)$$

This interaction has no real analogue in the filled-guide case, since it is a two-wave coupling near $\beta = \beta_e$. (We recall that all interactions near β_e in the filled-guide case were three-wave couplings.) The theory of weak coupling would predict that there should be traveling-wave amplification near $\omega = \omega_1$ (Figure 4.2), and backward-wave oscillations near $\omega = \omega_2$ (for every plasma mode q_n). These results can be verified by applying the analysis in Section 2.7.1 to Equation 4.72. The growth rates and starting lengths for these interactions can be computed in the same manner as that outlined in Section 4.1; that is, the frequencies ω_1 and ω_2 are solved for by setting $\beta_0 = \beta_e$, and these frequencies are then used to evaluate the right-hand side of Equation 4.72 (which determines the inter-action strength).

We note, in closing, that a comparison of the interaction strengths of the modes at n = ±1 at the synchronous frequencies with those away from synchronism (reactive-medium amplification) shows that the reactive-medium effects should become dominant for very small beam radius b, since all $r_n \to 0$ as $b \to 0$. The relative importance of the synchronous and nonsynchronous interactions in the case for n = ±1, then, is critically dependent on the beam diameter. This is in marked contrast to the case for n = 0, where the reactive-medium amplification rate almost always has its maximum at the synchronous frequencies.[30-32]

4.3 Interactions in an Infinite Magnetic Field

In the previous two sections we considered the interaction of an electron beam with a cold plasma in the presence of a finite steady magnetic field. In those sections it was necessary to make a num-ber of simplifying assumptions, such as very low density beams and/ or very thin beams, in order to obtain tractable results without de-tailed numerical computations. In this section it will be assumed that there is a large magnetic field which constrains the beam and plasma electrons to motion only along the field lines; however, the effects of large beam densities and more general beam geometries will be investigated.

From the general formulation in Section 4.1 it is clear that the assumption of an infinite magnetic field is reasonable only when $\omega_{ce} \gg \omega_{pe}$ and $\omega \gg \omega_{pi}$ in order that transverse plasma current be negligible ($K_\perp \simeq 1$). The second condition essentially requires that the frequency be high enough so that the plasma ions remain im-. mobile; if this is not the case, then the transverse motion of the ions should be accounted for. Therefore, this section will be con-cerned only with the interactions with the plasma electrons at high frequencies.

4.3.1 Filled Waveguide. We first consider the case where both the beam and the plasma fill a cylindrical waveguide structure (Fig-ure 4.1). The radial wave number is given by[28]

$$p^2 = -\left(\beta^2 - \frac{\omega^2}{c^2}\right)K_\parallel^0 \tag{4.73}$$

where $K_\parallel^0$ is defined in Section 4.1, and the eigenvalues of p^2 are determined by selecting the appropriate solutions of

$$\nabla^2\varphi + p^2\varphi = 0 \tag{4.74}$$

which match the boundary condition that $\varphi = 0$ on the waveguide
wall. Equation 4.73 is "dynamically" correct; that is, the quasi-
static assumption has not been made, and the function φ is proportional to the longitudinal electric field of an E-wave. For slow
waves ($\beta \gg \omega/c$), the quasi-static results are recovered, which
can also be obtained by setting $B_0 = \infty$ in Equation 4.3 ($K_\perp^0 = 1$).
For a circular geometry, which is of primary concern, the eigenvalues of p are given by $p = \epsilon_{nm}/b$, where ϵ_{nm} is the mth
root of the Bessel function $J_n(x)$.

The dispersion equation (4.73) can also be written in the form

$$\Delta(\omega, \beta) = 1 - \frac{\omega_{po}^2}{\omega^2} - \frac{\omega_{pb}^2}{(\omega - \beta v_0)^2} - \frac{p^2 c^2}{\omega^2 - \beta^2 c^2} = 0 \qquad (4.75)$$

This equation is of exactly the same <u>mathematical form</u> as that
obtained for one-dimensional longitudinal waves in a beam-plasma
system with cold ions and hot electrons, and with a square distribution for the electrons (see Equation 3.47). This system was
analyzed in some detail in Sections 3.3 through 3.5. The results
obtained for that system carry over directly to the present case
by the following interchange of symbols:

$$\left.\begin{array}{c} \omega_{pi} \rightarrow \omega_{po} \\[10pt] \omega_{pe} \rightarrow pc \\[10pt] v_{Te} \rightarrow c \end{array}\right\} \qquad (4.76)$$

(However, note that care must be exercised in carrying over results where approximations such as $m/M \ll 1$ have been made in
.the analysis in Chapter 3.)

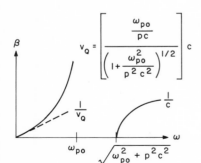

Figure 4.9. Plasma dispersion
for infinite magnetic field.

The plasma waves in the absence of the beam are shown in Figure 4.9; by analogy with Section 3.3, the condition for amplification

of the space-charge waves of a very <u>weak beam</u> is that

$$v_0 < v_Q = \left[\frac{\frac{\omega_{p0}}{pc}}{\left(1 + \frac{\omega_{p0}^2}{p^2 c^2}\right)^{\frac{1}{2}}} \right] c \tag{4.77}$$

which, for $v_0 \ll c$, requires that

$$\frac{\omega_{p0}}{pv_0} > 1 \tag{4.78}$$

We can also derive the general condition for an instability, following the approach in Section 3.4. If we assume that $\omega/\beta \ll c$, then the condition for an onset of complex ω for real β is that

$$\frac{\partial \Delta}{\partial \omega} = \frac{2\omega_{pb}^2}{(\omega - \beta v_0)^3} + \frac{2\omega_{p0}^2}{\omega^3} = 0 \tag{4.79}$$

for some real ω and real β that are also solutions of Equation 4.75. This requires that[28]

$$\left(\frac{\omega_{p0}^2}{p^2 v_0^2}\right) \left[1 + \left(\frac{\omega_{pb}^2}{\omega_{p0}^2}\right)^{\frac{1}{3}}\right]^3 \geq 1 \tag{4.80}$$

for an instability. It is interesting also to note that in the limit of $\omega_{p0} \to 0$ this condition becomes just $\beta_{pb} > p$. In Pierce's original analysis of the limiting stable current in neutralized beams,[69] he obtained a similar condition from a model in which he assumed an <u>immobile</u> ion plasma, which should correspond to $\omega_{p0} \to 0$ because the mass of the plasma particles is essentially infinite in this approximation.

We can also, by analogy with the results of Section 3.5, predict that an <u>infinite amplification rate</u> should set in at a critical value of the beam density. In the present case, this infinite amplification occurs when[28]

$$\beta_{pb}^2 > p^2 \tag{4.81}$$

A qualitative sketch of the dispersion for $\beta_{pb}^2 > p^2$ is presented in Figure 4.10. It can easily be verified that the complex roots of

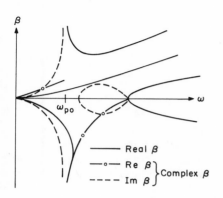

Figure 4.10. Beam-plasma dispersion for $\beta_{pb}^2 > p^2$.

β for real ω occur for $\omega < \omega_{po}$ rather than $\omega > \omega_{po}$ when β_{pb} becomes larger than p. In fact, a very careful expansion near $\omega \to \omega_{po}$ yields

$$\beta = \pm j\alpha + \left(\frac{\beta_{pb}^2}{\beta_{pb}^2 - p^2}\right)\beta_e \tag{4.82}$$

where

$$\alpha = \left(\frac{p^2 - \beta_{pb}^2}{K_{\parallel}}\right)^{\frac{1}{2}} \tag{4.83}$$

with $\alpha \to \infty$ (compare with Equation 3.50).

The total small-signal power flow in a single complex wave is zero; in the present case this power is composed of the beam kinetic power and the electromagnetic power. The kinetic power is proportional to $-\text{Re}\,(\beta - \beta_e)$; from Equation 4.82 we see that the kinetic power in the complex waves near ω_{po} becomes negative when β_{pb} becomes greater than p. This is not conclusive proof that the complex root of Equation 4.82 with $\beta_i > 0$ is an amplifying wave in this case; however, the numerical computations presented in Chapter 3 showed that this is indeed true (and also that there were no absolute instabilities present).

The condition for infinite amplification rate (Inequality 4.81) is a requirement on beam perveance alone in the case of filled circular guide, since

$$(\beta_{pb}b)^2 = \left(\frac{e\rho_{ob}}{\epsilon_0 m}\right)\left(\frac{1}{v_0^2}\right)\left(\frac{\pi b^2}{\pi}\right) = K\left[2\sqrt{2}\,\pi\epsilon_0\left(\frac{e}{m}\right)^{\frac{1}{2}}\right]^{-1} \simeq 3 \times 10^4 K \tag{4.84}$$

where the beam perveance is defined as

$$K = \frac{I_0}{V_0^{\frac{3}{2}}} \qquad\qquad (4.85)$$

and b is the radius of the beam and waveguide.

For the lowest-order mode, with pb = 2.405, it is necessary that the perveance be greater than 190×10^{-6} for the amplification rate to be infinite. In the following section the case of a beam that only partially fills the waveguide will be analyzed; in that case pb can be less than 2.405, and hence the critical perveance can be lower than 190×10^{-6}.

4.3.2 Unfilled Waveguide. In this section is discussed the extension of the condition for infinite amplification to cases in which the beam does not fill the waveguide.

We consider first the two-dimensional slab geometry shown in Figure 4.11a. The beam is of width 2b; the plasma fills the space

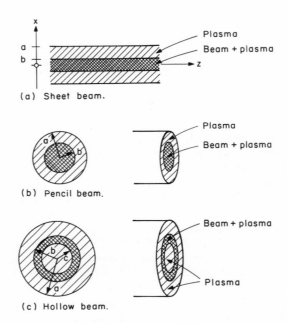

(a) Sheet beam.

(b) Pencil beam.

(c) Hollow beam.

Figure 4.11. Beam geometries.

between the two perfectly conducting planes at x = ±a. The quasi-static assumption can be made at the outset ($\beta \gg \omega/c$), since we shall be interested only in the limit of $|\beta| \to \infty$ as $\omega \to \omega_{po}$. The radial wave number p inside the beam region ($|x| < b$) is therefor

$$p^2 = -\beta^2 K_{\parallel}^0 = -\beta^2 \left[K_{\parallel} - \frac{\beta_{pb}^2}{(\beta - \beta_e)^2} \right] \tag{4.86}$$

The radial wave number outside the beam region is

$$q^2 = -\beta^2 K_{\parallel} \tag{4.87}$$

If we restrict our attention to the even modes, the potentials inside and outside the beam regions are

$$\varphi_i = \frac{\cos px}{\cos pb}, \qquad |x| < b \tag{4.88}$$

and

$$\varphi_0 = \frac{\sin q(a - x)}{\sin q(a - b)}, \qquad b < x < a \tag{4.89}$$

These potentials satisfy the boundary conditions that φ be zero at $x = a$ and that φ be continuous at $x = b$. Requiring that the x-component of the electric field be continuous at $x = b$, we obtain the following determinantal equation relating p and q:

$$pb \tan pb = qb \cot q(a - b) \tag{4.90}$$

When $p^2 < 0$ and real, Equation 4.90 takes the form

$$-|p| b \tanh |p| b = qb \cot q(a - b) \tag{4.91}$$

When $q^2 < 0$ and real, we have

$$pb \tan pb = |q| b \coth |q| (a - b) \tag{4.92}$$

It can easily be verified that no solution is possible for the case when both p^2 and q^2 are negative and real. By sketching both sides of Equations 4.90, 4.91, and 4.92, we can show that the general form of the relation between p and q for real values of p^2 and q^2 must be as shown in Figure 4.12.

From Equations 4.86 and 4.87, it follows that p and q must also be related by

$$p^2 = q^2 + \beta_{pb}^2 \left[\frac{\beta^2}{(\beta - \beta_e)^2} \right] \tag{4.93}$$

The expansion of β near the pole at ω_{po} carries through as in the case for the filled guide, except that now we evaluate $p^2(\beta)$ <u>at</u>

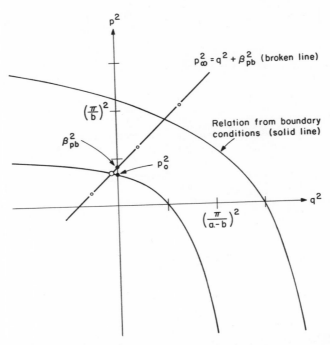

Figure 4.12. Sketch of p^2 - q^2 relations (illustrated for the case $\beta_{pb}^2 > p_0^2$).

$\beta = \infty$ in Equation 4.82, which we write as p_∞^2. We <u>assume</u> that the onset of infinite amplification is again signaled by a "flip" at the pole, a quite reasonable assumption because this is again the condition for this complex wave (with $\beta_i \to \infty$) to carry negative kinetic power, as was discussed in Section 4.3.1. It would seem plausible that the small-signal beam power <u>should</u> be negative for a wave to be amplifying, under the assumption that there are no absolute instabilities present, so that it makes sense to speak about real frequencies. This statement is as yet unproved. However, the author knows of no cases in which it is violated. That is, from Equation 4.82, an infinite amplification rate should be obtained when

$$\beta_{pb}^2 > p_\infty^2 \tag{4.94}$$

From Equation 4.93, the relation between p and q at $\beta = \infty$ is given by

$$p_\infty^2 = q^2 + \beta_{pb}^2 \tag{4.95}$$

This equation is also plotted in Figure 4.12. The intersections of Equation 4.95 with the curves derived from the determinantal equation (4.90) give the values of p^2 and q^2 at $\beta = \infty$. Because of the monotonically decreasing nature of p^2 vs. q^2, as shown in the figure, we see that the lowest-order mode will have $p^2_\infty < \beta^2_{pb}$ if and only if

$$\beta^2_{pb} > p^2_0 \qquad\qquad (4.96)$$

where p_0, the value of p obtained when $q = 0$ in the determinantal equation, is the solution of the transcendental equation

$$p_0 b \tan p_0 b = \frac{b}{a - b} \qquad\qquad (4.97)$$

For $a = b$, $p_0 b = \pi/2$, and the results of the filled-guide analysis are recovered. For $b \ll a$, however, $p_0 b$ is considerably less than $\pi/2$, and the condition for infinite amplification is much weaker.

It is interesting also to note that when $\beta^2_{pb} > p^2_0$ the value of q^2 at $\beta \to \infty$ for the lowest-order mode is negative; that is, when there is infinite amplification, the fields decay exponentially in the plasma region outside of the beam. This should be contrasted with the case when $\beta^2_{pb} < p^2_0$; then the fields have a <u>harmonic</u> variation in the plasma region outside of the beam, similar to the case of a plasma without the electron beam. Therefore, the fields of the complex wave at $\omega \to \omega_{po}$ are closely associated with the <u>beam</u> when $\beta_{pb} > p_0$, a result that is physically appealing. We also make the obvious observation that when $\beta_{pb} > \pi/b$ the higher-order modes can also have infinite amplification.

We now consider the cases of solid pencil beams and hollow beams, as shown in Figure 4.11b and c. Alexander[70] has considered the p^2 vs. q^2 relation for a solid pencil beam and has shown that it has the same general form as that depicted in Figure 4.12. For our purpose, the essential point about the p^2 vs. q^2 relation is that it has the monotonically decreasing form shown in Figure 4.12; that is, that $\partial p^2/\partial q^2$ be always negative. In Appendix I, Green's theorem is used to show that $\partial p^2/\partial q^2 < 0$ for any arbitrary shape of the beam. Therefore, it is true <u>in general</u> that there is infinite amplification if and only if $\beta_{pb} > p_0$, where p_0 is determined by setting $q = 0$ in the determinantal equation $D(p, q) = 0$ derived from the boundary conditions on the fields.

For the solid pencil beam, the determinantal equation for the $n = 0$ modes (see Equation 4.40) with $q = 0$ is

$$p_0 b \left[\frac{J_1(p_0 b)}{J_0(p_0 b)} \right] = \frac{1}{\ln\left(\dfrac{a}{b}\right)} \qquad\qquad (4.98)$$

Once again, for a = b the results for the filled guide are re-covered $[J_0(p_0 b) = 0]$. For a moderately thin beam (b ≪ a), the condition for infinite amplification (Inequality 4.96) becomes

$$\beta_{pb}^2 b^2 > \frac{2}{\ln\left(\dfrac{a}{b}\right)} \tag{4.99}$$

Using Equation 4.84, we can write Inequality 4.99 in terms of the beam perveance as

$$K > \frac{66}{\ln\left(\dfrac{a}{b}\right)} \times 10^{-6} \tag{4.100}$$

For a/b = 10, Inequality 4.100 requires that the perveance be greater than 29×10^{-6} for an infinite amplification rate.

For a hollow beam, the determinantal equation for the n = 0 modes with q = 0 is[70]

$$p_0 b \left[\frac{J_1(p_0 b) N_1(p_0 c) - J_1(p_0 c) N_1(p_0 b)}{J_0(p_0 b) N_1(p_0 c) - J_1(p_0 c) N_0(p_0 b)} \right] = \frac{1}{\ln\left(\dfrac{a}{b}\right)} \tag{4.101}$$

For a thin hollow beam, with b - c = δ and δ ≪ b, we can obtain more tractable results. Since

$$J_1(p_0 b) N_1(p_0 c) - N_1(p_0 b) J_1(p_0 c) \simeq p_0 \delta [J_0(p_0 b) N_1(p_0 b) - N_0(p_0 b) J_1(p_0 b)] \tag{4.102}$$

we find that

$$p_0^2 b \delta = \frac{1}{\ln\left(\dfrac{a}{b}\right)} \tag{4.103}$$

Therefore, the condition for infinite amplification is that

$$\beta_{pb}^2 b \delta > \frac{1}{\ln\left(\dfrac{a}{b}\right)} \tag{4.104}$$

which can also be written as

$$\left[\frac{1}{4\sqrt{2}\pi\epsilon_0 \left(\dfrac{e}{m}\right)^{\frac{1}{2}}} \right] \left[\frac{\rho_0 v_0 2\pi b \delta}{\left(\dfrac{m}{2e} v_0^2\right)^{\frac{3}{2}}} \right] > \frac{1}{\ln\left(\dfrac{a}{b}\right)} \tag{4.105}$$

or

$$K > \frac{66}{\ln\left(\frac{a}{b}\right)} \times 10^{-6}$$ (4.106)

in terms of the beam perveance.

It should be mentioned that the results derived here are directly applicable to cases in which the plasma does not fill the waveguide, if we take the radius a to be the radius of the plasma column. This follows from the fact that the radial wave number in a vacuum region is given by $q_{VAC}^2 = -\beta^2$; for this reason the fields outside of the plasma region would tend to zero as β tends to infinity, and it is therefore immaterial whether the plasma column is surrounded by a perfect conductor or by free space.

Chapter 5

INTERACTION WITH IONS IN A HOT-ELECTRON PLASMA

In Chapter 4 the interaction of an electron beam with a cold plasma in waveguide-type geometries was studied. In all of the configurations that were analyzed, it was generally true that the high-frequency interactions were stronger than those at the low frequencies characteristic of the plasma ions. The same sort of behavior was also indicated in the one-dimensional system analyzed in Chapter 3. There we found that a strong interaction with the ions results only when the plasma electrons are hot enough so that the beam space-charge wavelength is less than the plasma Debye wavelength. Physically, it is quite reasonable that the plasma electrons should be relatively hot for the beam to interact primarily with the ions, since "cool" electrons tend to "short out" a low-frequency electric field.

In this chapter we shall consider the case of a hot-electron plasma that fills a cylindrical waveguide structure. It will be assumed throughout that the ions are cold and that the electron beam is cold and also fills the waveguide. (The system has the same form as that shown in Figure 4.1.) The additional assumptions are listed now:

1. The quasi-static assumption will be made (see the introduction to Chapter 4). Since the region near ion plasma resonance is of primary interest and the beams of interest are nonrelativistic, this should not be too stringent an assumption.

2. It is assumed that there is a very large axial magnetic field applied that constrains the beam and plasma electrons to move only along the field lines. The transverse motion of the ions is allowed, however, because of their larger mass. This limits the applicability of the analysis to cases for which $\omega_{pe} \ll \omega_{ce}$.*

3. By virtue of the large magnetic field, only the longitudinal temperature of the plasma electrons is accounted for. This assumption is reasonable only if the Larmor radius V_{Te}/ω_{ce} of the electrons is much less than a transverse "wavelength," that is, if the magnetic field is large enough so that $pV_{Te}/\omega_{ce} \ll 1$.

4. It will usually be assumed that the plasma electrons are at a temperature much greater than the beam voltage ($T_e \gg V_0$), so that in the region of interest we can assume $\omega/\beta \ll V_{Te}$.

_____

*This assumption is not made in the analysis of the resistive-medium type of interaction.

Weak-Beam Interactions 119.

This analysis was initiated primarily to extend to the case of finite geometries the one-dimensional requirement $\lambda_{pb} < \lambda_{De}$ for interaction with ions as discussed in Section 3.5. It was found, however, that an absolute instability can also occur when the system has finite transverse bounds. As is shown in Section 5.2, this absolute instability occurs when the beam density is large enough so that ω_{pb} is (roughly) on the order of or larger than ω_{pi}, a very modest requirement. In Section 5.1 we analyze the waves on the plasma in the absence of the beam and the interaction with a very weak electron beam. This analysis provides some background for the results in Section 5.2.

5.1 Plasma Dispersion and Weak-Beam Interactions

In Appendix C the dispersion equation for a one-dimensional system was derived from a particle orbit or "multibeam" type of analysis. In this approach a single species of charged particles with some distribution of random velocities is considered to be composed of an infinite number of "beams," each with a <u>unique</u> zero-order velocity, that interact only through the <u>macroscopic</u> electromagnetic fields. This approach is completely equivalent to that of working with the collisionless Boltzmann-Vlasov equation.

In the present case, we have assumed that there is a very large magnetic field present, so that the unperturbed motion of the electrons consists of very tight spirals about the magnetic field (assumption 3). For this reason, the transverse variation of the electric field that is seen by an individual electron during its circular orbit is negligible, and only the "streaming" along the field lines is of importance. The plasma electrons, therefore, are equivalent in this case to a large number of electron beams in <u>confined flow</u>, with each beam having some unperturbed velocity along the magnetic field. From this physical picture, we can immediately write down the dispersion equation of the system by using the results of Section 4.1 and Appendix G. The radial wave number p is now given by

$$p^2 = -\beta^2 \frac{K_{\parallel}^0}{K_{\perp i}} \tag{5.1}$$

where

$$K_{\parallel}^0 = K_{\parallel i} - \frac{\omega_{pb}^2}{(\omega - \beta v_0)^2} - \omega_{pe}^2 \int_{-\infty}^{+\infty} \frac{f_{0e}(v_z)\, dv_z}{(\omega - \beta v_z)^2} \tag{5.2}$$

$$K_{\parallel} = 1 - \frac{\omega_{pi}^2}{\omega^2} \tag{5.3}$$

$$K_{\perp i} = 1 - \frac{\omega_{pi}^2}{\omega^2 - \omega_{ci}^2} \tag{5.4}$$

and $f_{0e}(v_z)$ is the plasma electron's velocity distribution. Note that since we are assuming that the electrons move only along the magnetic field, it is correct to replace $K_{\perp}^{0}$ by $K_{\perp i}$ as we have done in Equation 5.1 (see Equation 4.3). In the case of a filled waveguide, as considered here, p is completely determined from the cross-sectional geometry, as discussed in Section 4.1.

In the regime where $\omega \ll \beta V_{Te}$, the integration over v_z in Equation 5.2 takes the approximate form (see Equation 3.15)

$$\int_{-\infty}^{+\infty} \frac{f_{0e}(v_z)\, dv_z}{(\omega - \beta v_z)^2} \simeq - \frac{1}{\beta^2 V_{Te}^2} \tag{5.5}$$

for a Maxwellian distribution, where $eT_e/m = V_{Te}^2$ and T_e is the temperature of the plasma electrons in electron volts.

5.1.1 Plasma Dispersion. In the absence of the beam, the dispersion equation in the regime $\omega \ll \beta V_{Te}$ takes the form

$$\beta^2 = - \frac{\beta_{De}^2 + p^2 K_{\perp i}}{K_{\parallel i}} \tag{5.6}$$

where $\beta_{De} = \omega_{pe}/V_{Te}$ is the Debye wave number. This dispersion is sketched in Figure 5.1 for the case $\omega_{ci} < \omega_{pi}$.* When

$$\frac{\beta_{De}^2}{p^2} < \frac{\omega_{ci}^2}{\omega_{pi}^2 - \omega_{ci}^2} \tag{5.7}$$

the resonance at $\omega = \omega_{pi}$ belongs to a backward wave. We note that in the limit as $T_e \to \infty$, the dispersion is the same as that of a cold ion "cloud" in a waveguide, as considered by Smullin and Chorney.[1] This would be expected, since very hot electrons "see" a very high frequency field and therefore do not take part in the oscillations. For reasonable temperatures, however, Inequality 5.7 is reversed, and the plasma waves are forward waves (except for the very high order modes with large p^2).

This dispersion in a waveguide is quite different from that of purely longitudinal waves in a one-dimensional system, particularly in regard to the "cutoff" at $\omega = \omega_a$. This alteration of the

---

*Only the case of $\omega_{ci} < \omega_{pi}$ will be considered in detail, since this is essentially always the case for systems of interest.

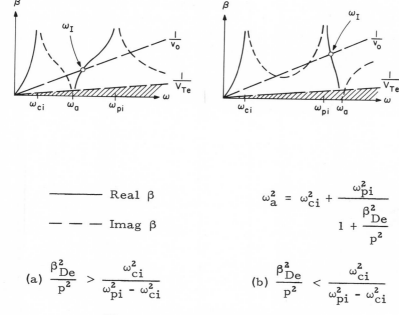

$$\omega_a^2 = \omega_{ci}^2 + \frac{\omega_{pi}^2}{1 + \frac{\beta_{De}^2}{p^2}}$$

(a) $\dfrac{\beta_{De}^2}{p^2} > \dfrac{\omega_{ci}^2}{\omega_{pi}^2 - \omega_{ci}^2}$ (b) $\dfrac{\beta_{De}^2}{p^2} < \dfrac{\omega_{ci}^2}{\omega_{pi}^2 - \omega_{ci}^2}$

Figure 5.1. Plasma dispersion.

dispersion caused by the boundaries is the result of a "coupling" of the one-dimensional transverse- and longitudinal-type waves. As will be shown in the following, there now can be reactive-medium amplification by a very weak beam even when v_0 is considerably greater than $(m/M)^{1/2}V_{Te}$ because of this alteration in the plasma dispersion.

5.1.2 Interaction with a Weak Beam. In the one-dimensional case analyzed in Section 3.3 it was found that a low-frequency reactive-medium instability arises only when $v_0 < (m/M)^{1/2}V_{Te}$ (for a very weak beam). For $v_0 \ll V_{Te}$, we can assume $\omega \ll \beta V_{Te}$ in Equation 5.1 and thus write the dispersion equation for the finite beam-plasma system in the form

$$\frac{\beta_{pb}^2}{(\beta - \beta_e)^2} = \frac{\beta_{De}^2 + p^2 K_{\perp i}}{\beta^2} + K_{\parallel i} = \frac{K_{\parallel i}}{\beta^2}[\beta^2 - \beta_0^2(\omega)] \qquad (5.8)$$

where $\beta_0(\omega)$ is the dispersion of the plasma wave in the absence of the beam, which is given by Equation 5.6. For a very weak beam, unless $\beta_e \simeq \beta_0$, we can evaluate the right-hand side of Equation 5.8 at $\beta = \beta_e$. We then find that reactive-medium amplification is obtained in the frequency range from ω_{ci} to ω_I, where ω_I is the frequency at which $\beta_0 = \beta_e$, as indicated in Figure 5.1.[*] When v_0 becomes comparable with V_{Te}, we would expect the Landau damping to come into play and cause this reactive-medium instability to pass over into a resistive-medium type of interaction, as will be discussed later in this section. When we are close to the synchronism with the plasma wave (where $\beta_0 \simeq \beta_e$), the dispersion equation (5.8) can be written as

$$(\beta - \beta_e)^2(\beta - \beta_0) = \frac{\beta_{pb}^2 \beta_e}{2K_{\parallel i}} \tag{5.9}$$

where the right-hand side can be evaluated at the intersection frequency ω_I for a very weak beam ($\beta_{pb} \to 0$). Equation 5.8 is very similar to the case for the cold plasma, Equation 4.14, except that $K_{\parallel}$ is now replaced by $K_{\parallel i}$. From the analysis in Appendix H, we can show that there is no absolute instability when the plasma wave is a forward wave, as in Figure 5.1a, and that the root with $\beta_i > 0$ is an amplifying wave in this case. (This dispersion equation is completely analogous to that of a traveling-wave tube with QC = 0.) When Inequality 5.7 is satisfied so that the plasma wave is a backward wave at synchronism, there is an absolute instability near the frequency ω_I with $\omega_I > \omega_{pi}$ (Figure 5.1b). (The dispersion equation is now analogous to that of a backward-wave oscillator with QC = 0.) The computation of the amplification rates and starting lengths can be carried through in exactly the same manner as in Section 4.1. In the present case the C parameter (for the case illustrated in 5.1a or 5.1b) is given by

$$C^3 = \left| \frac{\beta_{pb}^2}{2\beta_e^2 K_{\parallel i}} \right| \tag{5.10}$$

---

[*]Note that there can also be an interaction below ω_{ci} if the plasma wave has a phase velocity at $\omega \to 0$ which is greater than v_0. For this to occur, we must have

$$\frac{p^2 v_0^2}{\omega_{pi}^2}\left(1 + \frac{\omega_{pi}^2}{\omega_{ci}^2}\right) + \frac{\omega_{pe}^2}{\omega_{pi}^2} \frac{v_0^2}{V_{Te}^2} < 1$$

which requires (at least) that V_0 be less than $\frac{1}{2}(m/M)T_e$. For this reason, this interaction is of less physical importance and will not be considered in this section (see, however, the computations illustrated in Figure 5.8).

where it is to be evaluated at the intersection frequency ω_I. As an example, consider the maximum growth rate of the amplifying wave (Figure 5.1a) when $\omega_{pi} \gg \omega_{ci}$, so that

$$\omega_I^2 \simeq \frac{\omega_{pi}^2}{1 + \frac{\beta_{De}^2}{p^2}} \qquad (5.11)$$

This maximum growth rate is

$$(\text{Im } \beta)_{max} = \frac{\sqrt{3}}{2^{\frac{4}{3}}} \beta_{pb} \left(\frac{\omega_{pi}}{\omega_{pb}}\right)^{\frac{1}{3}} \frac{\frac{p}{\beta_{De}}}{\left(1 + \frac{p^2}{\beta_{De}^2}\right)^{\frac{7}{6}}} \qquad (5.12)$$

Equation 5.12 illustrates the general rule that the weak-beam assumption is valid only when $\omega_{pb} \ll \omega_{pi}$.

    5.1.3 Resistive-Medium Amplification. In the one-dimensional case, we found that a strong resistive-medium amplification at high frequencies arises when v_0 is comparable with V_{Te}. The physical mechanism for the loss is the Landau damping. To investigate the resistive-medium amplification in the present case, the Im β has been computed using a resonance distribution for the plasma electrons. The model of the plasma is slightly different in the present case: Only the longitudinal temperature of the plasma electrons is accounted for; however, small-signal transverse motion of the plasma electrons is allowed, and the assumption of $\omega \ll \beta V_{Te}$ is not made. The resulting dispersion equation is developed in Appendix J. When a very weak beam is assumed, the solution for the space-charge waves takes the form

$$\frac{\beta_{pb}^2}{(\beta - \beta_e)^2} = 1 - \frac{\omega_{pi}^2}{\omega^2} - \frac{\omega_{pe}^2}{\omega^2 \left(1 - j\frac{V_T}{v_0}\right)^2}$$

$$+ \frac{p^2 v_0^2}{\omega^2} \left[1 - \frac{\omega_{pi}^2}{\omega^2 - \omega_{ci}^2} - \frac{\omega_{pe}^2}{\omega^2 \left(1 - j\frac{V_T}{v_0}\right)^2 - \omega_{ce}^2}\right]$$

$$(5.13)$$

The amplification rate obtained from Equation 5.13 is plotted as a function of frequency in Figure 5.2 for an electron-proton plasma with $\omega_{pe} = \omega_{ce}$, $pv_0/\omega_{pe} = 0.1$, and $V_{Te}/v_0 = 1$ and 5. The amplification curve for $V_{Te} = v_0$ is roughly of the same form as

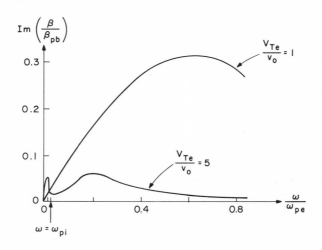

Figure 5.2. Resistive-medium amplification.

the one-dimensional curve (see Figure 3.4). However, for $V_{Te} = 5v_0$ we note that there is a low-frequency peak developing below ω_{pi}. This low-frequency peak is caused by a "synchronism" of the beam with a (damped) plasma wave near ω_I; as V_{Te} becomes much larger than v_0, the "effective dielectric constant" of the plasma becomes largely inductive rather than resistive-inductive, and this low-frequency resistive-medium peak passes over into a reactive-medium amplification. (The exact transition point between the resistive and reactive mechanisms depends on beam density as well as V_{Te}/v_0.)

5.2 Strong Interaction with Ions

In Section 5.1, the low-frequency interactions in a hot-electron plasma were analyzed for the case in which the electron beam has a vanishingly small density ($n_b \to 0$). In the present section, these interactions will be investigated for cases of stronger electron beams. It will be assumed initially that $v_0 \ll V_{Te}$, so that the dispersion in the limit of $\omega \ll \beta V_{Te}$ will cover the range of interest. Some results of computations by Puri[71] will be presented in Section 5.2.4 to illustrate the transition to lower temperature.

The present study was initiated in order to extend the one-dimensional results of Section 3.5 to the case of finite systems. However, in the course of this investigation it was found that an absolute instability can arise at a critical (and quite reasonable) value of beam density even when the plasma waves are all forward waves. (This occurs roughly when ω_{pb} becomes comparable with ω_{pi}.) In addition, it is still true that infinite amplification can be obtained when λ_{pb} is (roughly) less than λ_{De}, as in the one-dimensional case. (The meaning of this amplification becomes muddled, however, when the absolute instability is present at the same time.)

In the following, the condition for the appearance of an absolute instability is discussed first, and then the condition for the appearance of infinite amplification is formulated.

5.2.1 <u>Absolute Instability</u>. If we assume that $\omega \ll \beta V_{Te}$, the dispersion equation can be written as

$$1 - \frac{\omega_{pi}^2}{\omega^2} + \frac{\beta_{De}^2 + p^2 K_{\perp i}}{\beta^2} - \frac{\omega_{pb}^2}{(\omega - \beta v_0)^2} = 0 \qquad (5.14)$$

from Equation 5.1 through 5.5. This equation is fourth order in β and sixth order in ω. In Figures 5.3 and 5.4 the plot of complex β for real values of ω is presented for the case of $p^2/\beta_{pb}^2 = 100$, $\beta_{De}^2/\beta_{pb}^2 = 100$, $\omega_{ci}/\omega_{pi} = 0.05$, and for $\omega_{pi}/\omega_{pb} = 2$ and 0.75. The very low frequency region near ω_{ci} is not shown on these plots. Also note that the dispersion of the plasma wave in the absence of the beam is as shown in Figure 5.1a for these numbers. The corresponding plots of the loci of (two of) the roots of β for complex ω are presented in Figures 5.5 and 5.6, illustrating the application of the criteria of Chapter 2.

In the case of $\omega_{pi}/\omega_{pb} = 2$, there is no absolute instability, and one of the roots of complex β for <u>real</u> ω with $\beta_i > 0$ is an amplifying wave, as can be seen from Figure 5.5. (Actually, in any region where <u>all</u> roots of β for <u>real</u> ω are complex, it must be true that <u>one</u> root with $\beta_i > 0$ is an amplifying wave. This follows from the fact that <u>three</u> roots of β are in the lower-half β-plane for $\omega_i \to -\infty$, as can be easily verified, and therefore one root with $\beta_i > 0$ for real ω must be a "downstream" wave.) The complex root that is amplifying corresponds to the "beam wave" with $\beta_r \simeq \beta_e$, as we would expect from the analysis of the weak-beam interactions.

In the case of $\omega_{pi}/\omega_{pb} = 0.75$, however, we find that there <u>is</u> an absolute instability because two roots of β that come from different halves of the complex β-plane for $\omega_i \to -\infty$ merge into a double root at the frequency $\omega_s = \omega_{pi}(0.707 - j0.001)$ (Figure 5.6). Note also that there is an abrupt "switch" of the roots of complex β for real ω from amplifying to evanescent in this case (see the discussion of this general behavior in Sections 2.5 and 2.6). That is, the complex root with the larger β_i becomes a "downstream" wave for $\omega_r > 0.707\omega_{pi}$ (see Figures 5.4 and 5.6).

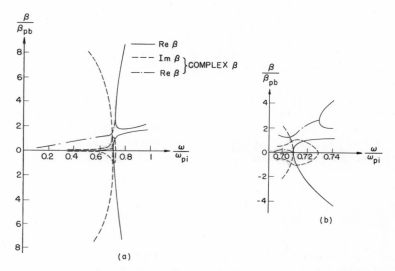

Figure 5.3. (a) Dispersion for $\omega_{pi}/\omega_{pb} = 2$. (b) Detail near $\omega = 0.7\omega_{pi}$.

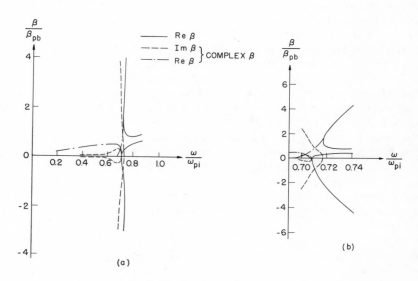

Figure 5.4. (a) Dispersion for $\omega_{pi}/\omega_{pb} = 0.75$. (b) Detail near $\omega = 0.7\omega_{pi}$.

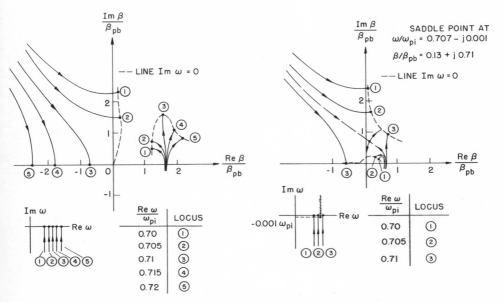

Figure 5.5. Locus of β for complex ω, $\omega_{pi}/\omega_{pb} = 2$.

Figure 5.6. Locus of β for complex ω, $\omega_{pi}/\omega_{pb} = 0.75$.

If we look at the plots of complex β for <u>real</u> ω in both cases (Figures 5.3 and 5.4), we note that there is an "interchange" of where the complex roots end. That is, the "cutoff plasma wave" that has $\beta_i \to \infty$ at $\omega = \omega_{ci}$ (see also Figure 5.1a) ends at the lower branch point of $\beta(\omega)$ ($\partial\beta/\partial\omega = \infty$ for real ω) for the case $\omega_{pi}/\omega_{pb} = 2$, and at the upper branch point for $\omega_{pi}/\omega_{pb} = 0.75$. Correspondingly, the "<u>beam waves</u>" terminate at the upper branch point for the lower beam density $\omega_{pi}/\omega_{pb} = 2$, as would be the case for a very weak beam, and at the lower branch point for the larger beam density $\omega_{pi}/\omega_{pb} = 0.75$.

These computations illustrate the important transition that takes place when the beam density (relative to the plasma density) exceeds a critical value. This transition in terms of the plot of complex β for real ω is illustrated again in Figure 5.7. The case of a weak beam is shown in Figure 5.7a; as we have demonstrated in Section 5.1, and verified by the computations presented in Figure 5.5, there is no absolute instability in this case. The computations for the case $\omega_{pi}/\omega_{pb} = 0.75$, however, have shown that there is an absolute instability for this value of beam density, and that the corresponding plot of β for real ω is of the form illustrated in Figure 5.7c. It is clear that there must be a critical set of parameters for which the plot is of the form illustrated in Figure 5.7b. That is, there must be a degeneracy (double root) of <u>complex</u> β for some real frequency when the parameters have their critical values in order that the complex

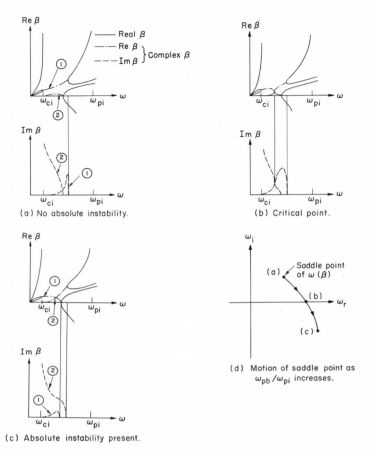

(a) No absolute instability.

(b) Critical point.

(c) Absolute instability present.

(d) Motion of saddle point as ω_{pb}/ω_{pi} increases.

Figure 5.7. Illustration of transition.

roots can "cross over" from the case of Figure 5.7a to Figure 5.7c. Stated in another way, when the parameters take on their critical values, a saddle point of $\omega(\beta)$ is in the process of crossing the real-ω axis and passing into the lower-half ω-plane (Figure 5.7d).

The preceding interpretation has been verified for the particular numerical case discussed. For example, in the case of $\omega_{pi}/\omega_{pb} = 2$, as shown in Figure 5.5, there is a saddle point of $\omega(\beta)$ at the frequency $\omega_s = \omega_{pi}(0.707 + j0.0015)$ and at the wave number $\beta_s = \beta_{pb}(0.84 + j1.0)$. The formation of this saddle point is almost obvious from the loci shown in Figure 5.5, even though the roots of β are not shown for complex ω in the upper-half plane. This saddle point has passed into the lower-half plane when $\omega_{pi}/\omega_{pb} = 0.75$ (Figure 5.6).

If we <u>postulate</u> that this double root of β for real ω at a complex β represents the onset of an absolute instability, we can solve for the critical parameter values at which it occurs. This procedure is carried through in Appendix K. The condition obtained there is that

$$\frac{\omega_{pb}^2}{\omega_{pi}^2} \geq 1 - \frac{\omega_{ci}^2}{\omega_{pi}^2} - \frac{p^2}{p^2 + \beta_{De}^2 + \beta_{pb}^2} \qquad (5.15)$$

for an absolute instability. This condition agrees with the numerical results obtained before, since for those numbers it required that $\omega_{pi}/\omega_{pb} < \sqrt{2}$ for an absolute instability (which was, in fact, the reason for choosing these two cases).

As is pointed out in Appendix K, Condition 5.15 is valid only when $p^2 \neq 0$. If $p^2 = 0$, we have the case of longitudinal waves in a one-dimensional system, and it is shown in the appendix that there is no such transition. This agrees with the analysis of Chapter 3, which shows that there is no absolute instability of the longitudinal wave in a one-dimensional system.

If β_{pb}^2 is much less than p^2 and β_{De}^2, the right-hand side of Equation 5.15 becomes <u>negative</u> when

$$\beta_{De}^2 < p^2 \left(\frac{\omega_{ci}^2}{\omega_{pi}^2 - \omega_{ci}^2} \right) \qquad (5.16)$$

Inequality 5.16 is exactly the condition for the plasma in the absence of the beam to support a backward wave (see Inequality 5.7). It was shown in Section 5.12 that there is an absolute instability of a very weak beam when Inequality 5.16 is satisfied, and is in agreement with the general condition in Equation 5.15.

We also note in passing that when $\omega_{ci} > \omega_{pi}$ the right-hand side of Equation 5.15 is always negative. From Equation 5.6, it can be shown that there is always a backward wave on the plasma (in the absence of the beam) when $\omega_{ci} > \omega_{pi}$, and therefore we should have an absolute instability in this case even when $n_b \rightarrow 0$. This is again in agreement with Equation 5.15.

5.2.2 Infinite Amplification at ω_{pi}. In the one-dimensional system analyzed in Chapter 3 it was found that an infinite amplification rate at ω_{pi} can arise when $\beta_{pb} > \beta_{De}$, or $\lambda_{pb} < \lambda_{De}$ (for cold ions, naturally). The onset of this infinite amplification could, in that case, be predicted by noting that there was a "flip" from complex to real β on one side of the pole at $\omega = \omega_{pi}$ (see Section 3.5). In addition, the condition for the complex wave to carry negative kinetic power was shown to agree with the amplification criteria of Chapter 2 for the cases computed.

In the present case, the expansion of $|\beta| \rightarrow \infty$ as $\omega \rightarrow \omega_{pi}$ takes the form

$$\beta^2 \simeq \frac{\beta_{pb}^2 + p^2\left(\dfrac{\omega_{ci}^2}{\omega_{pi}^2 - \omega_{ci}^2}\right) - \beta_{De}^2}{1 - \dfrac{\omega_{pi}^2}{\omega^2}} \tag{5.17}$$

therefore, for

$$\beta_{pb}^2 + p^2\left(\frac{\omega_{ci}^2}{\omega_{pi}^2 - \omega_{ci}^2}\right) > \beta_{De}^2 \tag{5.18}$$

the complex root of β with $\beta_i \to \infty$ as $\omega \to \omega_{pi}$ is obtained for $\omega < \omega_{pi}$. A more careful expansion of the dispersion equation (5.8) shows that

$$\beta \simeq \pm j\alpha + \left[\frac{\beta_{pb}^2}{\beta_{pb}^2 - \beta_{De}^2 + p^2\left(\dfrac{\omega_{ci}^2}{\omega_{pi}^2 - \omega_{ci}^2}\right)}\right]\beta_e \tag{5.19}$$

with

$$\alpha^2 = -\frac{\beta_{pb}^2 + p^2\left(\dfrac{\omega_{ci}^2}{\omega_{pi}^2 - \omega_{ci}^2}\right) - \beta_{De}^2}{1 - \dfrac{\omega_{pi}^2}{\omega^2}} \tag{5.20}$$

as $\omega \to \omega_{pi}$ (with $\omega < \omega_{pi}$ when Inequality 5.18 is satisfied). Since the beam kinetic power has the sign of $-\text{Re}\,(\beta - \beta_e)$, we find that this complex wave with $\beta_i \to \infty$ has negative kinetic power when

$$\beta_{pb}^2 + p^2\left(\frac{\omega_{ci}^2}{\omega_{pi}^2 - \omega_{ci}^2}\right) > \beta_{De}^2 > p^2\left(\frac{\omega_{ci}^2}{\omega_{pi}^2 - \omega_{ci}^2}\right) \tag{5.21}$$

When $\beta_{De}^2 < R^2\omega_{ci}^2/(\omega_{pi}^2 - \omega_{ci}^2)$, however, we know from Equation 5.15 that an absolute instability must be present for any value of of ω_{pb}/ω_{pi}. We would not expect arguments based on the concept of sinusoidal steady-state power flow to have any validity when there is an absolute instability present, so that the only criterion which should have any meaning in regard to infinite amplification is that given by Inequality 5.18 (left-hand side of Inequality 5.21). Note carefully, however, that these arguments based on kinetic power can be considered only as a conjecture, since it has not been <u>proved</u> that an amplifying wave must carry negative kinetic power. The plot of complex β for real ω for the parameter values $p^2 = \beta_{pb}^2 = \beta_{De}^2$, $\omega_{ci} = \frac{1}{2}\omega_{pi}$, and $\omega_{pi}/\omega_{pb} = 3$ is presented in Figure 5.8. These parameters are not reasonable ones from a practical standpoint; they were chosen only to illustrate a case where there should not be an absolute instability, according to Equation 5.15, while at the same time, the condition on kinetic power would suggest that there <u>is</u> infinite amplification. The application of the instability criteria for this case is illustrated in Figure 5.9; we see that both of our expectations are borne out.

Note that this (postulated) condition for infinite amplification in finite geometries as given by Inequality 5.21 is <u>less stringent</u> than it is in a one-dimensional system (when $\omega_{pi} > \omega_{ci}$); although in most cases it is essentially the same since we usually have ω_{ci} <u>much</u> less than ω_{pi}. Under most circumstances, however, the beam densities that would be required to make $\beta_{pb} > \beta_{De}$ would almost certainly result in an absolute instability. In order that there be no absolute instability, it is necessary that $n_b/n_p < m/M$; If β_{pb} is at the same time larger than β_{De}, then the plasma temperature must be larger than $2(M/m)V_0$, where V_0 is the dc beam voltage.

A case where Inequality 5.18 is satisfied at the same time that an absolute instability is present is shown in Figures 5.10 and 5.11 (parameters $p^2 = \beta_{De}^2 = \beta_{pb}^2$, $\omega_{ci} = \frac{1}{2}\omega_{pi}$, and $\omega_{pi}/\omega_{pb} = 0.25$). The frequency of the absolute instability is now $\omega_s \simeq (0.925 - j0.17)\omega_{pi}$, and the wave number where the saddle point occurs is estimated as $\beta_s \simeq \beta_{pb}(0.2 + j0.8)$ (Figure 5.11). Once again, there is an abrupt "switch" of the complex β for real ω from "evanescent" to "amplifying." It has been verified (although this range of ω is not plotted in Figure 5.11) that the smaller complex β for real ω (with $\beta_i > 0$) is amplifying over its entire range. The <u>larger</u> β_i becomes an "amplifying wave" for $\omega_r \gtrsim 0.925\omega_{pi}$, and the amplification rate tends to infinity as $\omega \to \omega_{pi}$, as can be seen from Figure 5.11. (The description of these waves as "amplifying" when there is an absolute instability present is subject to the qualifications described in Section 2.6.4.)

The case of $T_e = \infty$, $p^2 = \beta_{pb}^2$, $\omega_{ci} = \frac{1}{2}\omega_{pi}$, and $\omega_{pi}/\omega_{pb} = 0.25$ is presented in Figures 5.12 and 5.13. For this case the plasma in the absence of the beam supports a backward wave (see Inequality 5.7), and an absolute instability should be present according to Equation 5.15. From Figure 5.13 we see that an absolute instability is obtained near $\omega \sim \omega_{pi}(0.92 - j0.43)$ and $\beta \sim \beta_{pb}(0.4 + j0.5)$. It is interesting to note that the wave with $\beta_i \to \infty$ as $\omega \to \omega_{pi}$ is again an "amplifying wave," even though this complex wave carries positive kinetic power in this case (see Inequality 5.21).

As a matter of interest, the plot of complex β for real ω is

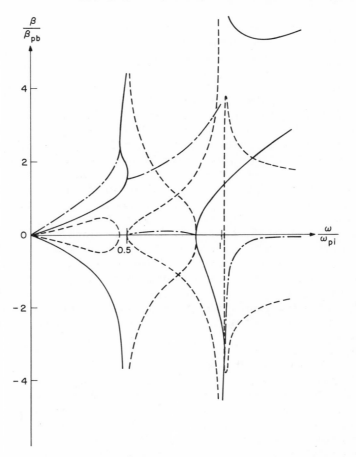

Figure 5.8. Dispersion for $p^2 = \beta_{De}^2 = \beta_{pb}^2$, $\omega_{ci} = \frac{1}{2}\omega_{pi}$, and $\omega_{pi}/\omega_{pb} = 3$.

presented in Figure 5.14 for the case $\beta_{De}^2 = 2\beta_{pb}^2$, $p^2 = \beta_{pb}^2$, $\omega_{ci} = \frac{1}{2}\omega_{pi}$, and $\omega_{pi}/\omega_{pb} = 0.25$. Complex β is obtained for $\omega > \omega_{pi}$ in this case, in agreement with Inequality 5.18.

   5.2.3 Summary of Results. From the preceding discussions, two general rules may be conjectured:

   1. An absolute instability is obtained when Inequality in Condition 5.15 is satisfied.

   2. Infinite amplification at ω_{pi} is obtained when Inequality 5.18 is satisfied.

   These two conditions are essentially independent of one another. That is, parameter values can be found for which either, both, or none of these conditions is satisfied. These general rules were found to be valid for every case that was computed.

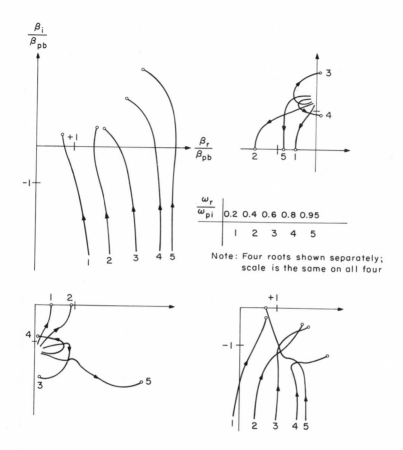

Figure 5.9. Stability criteria for $p^2 = \beta_{De}^2 = \beta_{pb}^2$, $\omega_{ci} = \frac{1}{2}\omega_{pi}$, and $\omega_{pi}/\omega_{pb} = 3$.

5.2.4 Extension to Lower Temperatures.

The preceding results were derived from an analysis of a dispersion equation valid only in the limit $\omega/\beta \ll V_{Te}$ ($T_e \gg V_0$). With this approximation, we found that an absolute instability could arise for very modest values of beam density. It is therefore not clear from the preceding work that it is <u>necessary</u> to have $T_e \gg V_0$ in order to have an absolute instability for physically reasonable beam densities.

An analysis of the dispersion equation (5.1) using the Maxwellian distribution without this approximation would be a difficult computational task. Puri[71] has performed computations on the dispersion equation (5.1) using a square distribution for $f_{0e}(v_z)$. This dispersion equation is

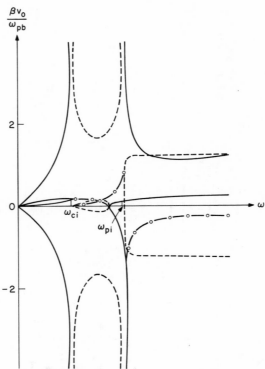

Figure 5.10. Dispersion for $p^2 = \beta_{De}^2 = \beta_{pb}^2$, $\omega_{ci} = \frac{1}{2}\omega_{pi}$, and $\omega_{pi}/\omega_{pb} = 0.25$.

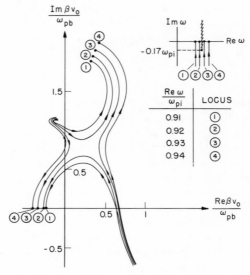

Figure 5.11. Stability criteria for $p^2 = \beta_{De}^2 = \beta_{pb}^2$, $\omega_{ci} = \frac{1}{2}\omega_{pi}$, and $\omega_{pi}/\omega_{pb} = 0.25$.

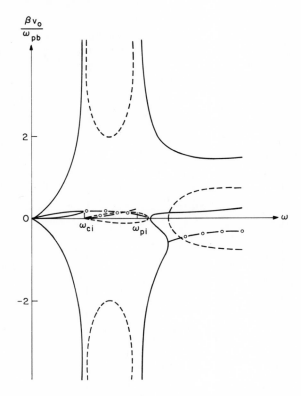

Figure 5.12. Dispersion for $T_e = \infty$, $p^2 = \beta_{pb}^2$, $\omega_{ci} = \frac{1}{2}\omega_{pi}$, and $\omega_{pi}/\omega_{pb} = 0.25$.

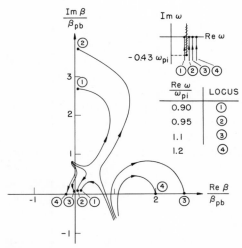

$\dfrac{\text{Re }\omega}{\omega_{pi}}$	LOCUS
0.90	①
0.95	②
1.1	③
1.2	④

Figure 5.13. Stability criteria for $T_e = \infty$, $p^2 = \beta_{pb}^2$, $\omega_{ci} = \frac{1}{2}\omega_{pi}$, and $\omega_{pi}/\omega_{pb} = 0.25$.

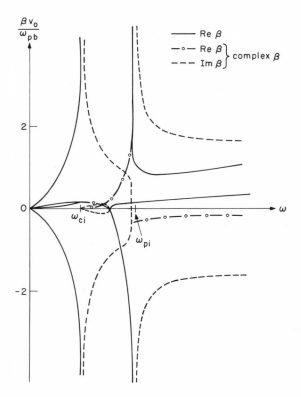

Figure 5.14. Dispersion for $\beta^2_{De} = 2\beta^2_{pb}$, $p^2 = \beta^2_{pb}$, $\omega_{ci} = \frac{1}{2}\omega_{pi}$, and $\omega_{pi}/\omega_{pb} = 0.25$.

$$1 - \frac{\omega^2_{pi}}{\omega^2} - \frac{\omega^2_{pb}}{(\omega - \beta v_0)^2} + \frac{p^2 K_{\perp i}}{\beta^2} - \frac{\omega^2_{pe}}{\omega^2 - \beta^2 V^2_{Te}} = 0 \qquad (5.22)$$

(see Equations 3.9 and 3.12).

In Figure 5.15, the critical beam density for an absolute instability to occur is plotted as a function of V^2_{Te}/v^2_0, with the other parameters fixed at the values given in the figure. The absolute instability occurs whenever the beam density exceeds the value indicated in the figure. For comparison purposes, the result predicted by Equation 5.15, which assumes $v_0 \ll V_{Te}$, is also illustrated. We see that the absolute instability can be obtained for values of V_{Te} comparable with v_0; however, the necessary beam density is increased correspondingly.

As a matter of interest, the spatial growth rate of the amplifying wave (in cases where there is no absolute instability) is plotted as a function of frequency in Figure 5.16 for several values of

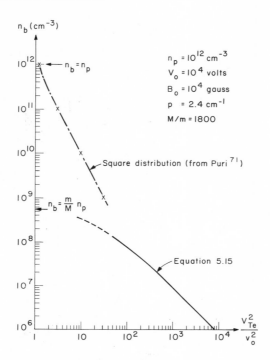

Figure 5.15. Relation between beam density n_b and temperature V_{Te}^2/v_0^2 for onset of absolute instability. The lower line is the relation given by Equation 5.15, and the upper line is the result of computations in Equation 5.22.[71]

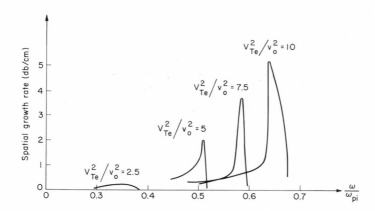

Figure 5.16. Spatial growth rate versus frequency for several values of temperature (from Puri[71]). Beam density is 10^{10} cm^{-3}, and other parameters are the same as given in Figure 5.15.

V_{Te}/v_0. The beam density is 10^{10} cm^{-3} in all cases, and the other parameters are the same as given in Figure 5.15. We note the rapid increase of the interaction strength as V_{Te}/v_0 becomes substantially greater than unity.

Appendix A

COMPARISON WITH THE METHOD OF FAINBERG, KURILKO, AND SHAPIRO

In Reference 51 the distinction between absolute and convective instabilities was investigated by determining the asymptotic behavior of a disturbance that is in the form of a spatial pulse at $t = 0$. Therefore, for later times this response can be written as[51]

$$\psi(t, z) = \sum_{\alpha} \int_{-\infty}^{+\infty} h_{\alpha}(k) e^{j[\omega_{\alpha}(k)t - kz]} \frac{dk}{2\pi} \tag{A.1}$$

where $\omega_{\alpha}(k)$ is one branch (or sheet) of the dispersion relation $\omega = \omega(k)$. (See also Polovin.[52]) Since $\psi(t = 0, z)$ is a pulse, these authors implicitly assume that $h_{\alpha}(k)$ is an entire function of k, so that the limiting value of the integral over k as $t \to \infty$ is determined by

$$\lim_{t \to \infty} I_{\alpha} \equiv \lim_{t \to \infty} \int_{-\infty}^{+\infty} e^{j\omega_{\alpha}(k)t} \frac{dk}{2\pi} \tag{A.2}$$

as was originally suggested by Landau and Liftshitz.[45]

In this approach, therefore, they assume that the various sheets of $\omega(k)$ can be handled independently. By changing the variable of integration, Equation A.2 can be written as

$$I_{\alpha} = \int_{\Omega_{\alpha}} \frac{e^{j\omega_{\alpha}t}}{\dfrac{d\omega_{\alpha}}{dk}} \frac{d\omega_{\alpha}}{2\pi} \tag{A.3}$$

where Ω_{α} is the contour of complex ω for real k as k runs from $-\infty$ to $+\infty$. From this form it is concluded that if there is a saddle point of $\omega_{\alpha}(k)$ ($d\omega_{\alpha}/dk = 0$) inside of the Ω_{α} contour and the real-ω axis, the integral diverges as $t \to \infty$, indicating an absolute instability (Figure 2.19).

The formalism developed in Sections 2.2 and 2.3 can be easily adapted to generate the same physical situation assumed in Reference 51, with an added advantage that the excitation of the various

139

normal modes can be explicitly considered. Assume a spatially
localized source, as before, but now let f(t) correspond to a pulse
in time which is nonzero only over the interval $(-T, 0)$. At the
instant $t = 0^+$, the response will be in the form of a spatial pulse,
and the subsequent history of this pulse on the undriven system
can then be determined. (See the discussion of this case in Section 2.3.5.) To correspond to the form used in Reference 51, we
shall now write the response in the form

$$\psi(t, z) = \int_{-\infty}^{+\infty} F(t, k)g(k)^{-jkz} \frac{dk}{2\pi} \tag{A.4}$$

where

$$F(t, k) \equiv \int_{-\infty-j\sigma}^{+\infty-j\sigma} G(\omega, k)f(\omega)e^{j\omega t} \frac{d\omega}{2\pi} \tag{A.5}$$

Notice that both $g(k)$ and $f(\omega)$ are now entire functions, since the
source is a pulse in both space and time. For $t > 0$, the integration in Equation A.5 can be closed in the upper-half ω-plane, and
$F(t, k)$ can thus be written as a sum over all of the normal modes,
that is, the solutions $\omega_\alpha(k)$ for a particular choice of k, since the
integration in Equation A.5 passes <u>below</u> all the poles of $G(\omega, k)$ in
the ω-plane. Mathematically stated,

$$F(t, k) = \sum_\alpha \frac{jf[\omega_\alpha(k)]}{\left(\dfrac{\partial G^{-1}}{\partial \omega}\right)_{\omega=\omega_\alpha(k)}} e^{j\omega_\alpha(k)t} \tag{A.6}$$

where $\omega_\alpha(k)$ are <u>all</u> of the solutions of $G^{-1}(\omega, k) = 0$ $[\Delta(\omega, k) = 0]$
for that particular k. Therefore, comparing Equation A.1 with
Equations A.6 and A.4, we see that in our formulation the function $h_\alpha(k)$ in Equation A.1 is

$$h_\alpha(k) = \frac{jf[\omega_\alpha(k)]g(k)}{\left(\dfrac{\partial G^{-1}}{\partial \omega}\right)_{\omega=\omega_\alpha(k)}} \tag{A.7}$$

The function $h_\alpha(k)$ is therefore not an entire function of k but
has singularities, which are actually branch poles, at the branch

points of $\omega(k)(d\omega/dk = \infty)$, where $\partial G^{-1}/\partial\omega \sim \partial\Delta/\partial\omega = 0$. Note, however, that the sum over α of all such $h_\alpha(k)$ does <u>not</u> have any singularities; that is, the "total" function $F(t,k)$ <u>is</u> an entire function of k. This must be so since the branch points of $\omega(k)$ are just double roots of ω for some k, and <u>both roots</u> are <u>always</u> included in the residue evaluation of Equation A.6, thus causing a cancellation between the two terms that individually have a singularity. It is also quite apparent physically that $F(t,k)$ must be an entire function of k, since $\psi(t,z)$ (Equation A.4) must be spatially bounded for all finite t.

In brief, the essential argument with the derivation given in Reference 51 is that the various sheets of $\omega(k)$ <u>cannot</u> always be handled separately. The points where the sheets "join," that is, the branch points of $\omega(k)$, give rise to singularities in the individual functions $h_\alpha(k)$ which must be considered. There is nothing wrong in principle with continuing the analysis in the form given by Equation A.4 and A.6; however, it is far less complicated to use the reverse-integration procedure carried through in Equations 2.15 and 2.16 of Section 2.3.

To illustrate a specific case in which the two criteria do not agree, we will consider a somewhat artificial numerical example with the dispersion equation

$$(\omega - k)^2 = 1 - 4j\omega \tag{A.8}$$

Consider Equation A.8 to be an approximation to some more exact dispersion equation where the approximations are valid only for reasonably small ω and k. The loci of the roots of k as ω_i is varied are sketched in Figure A.1 for $\omega_r = 0$. Since both roots of k are in the lower-half plane for sufficiently large negative imaginary part of ω, the saddle point does <u>not</u> represent an absolute instability. However, it might be anticipated that the criterion of Fainberg, Kurilko, and Shapiro would give the opposite result because one of the merging roots of k has crossed the real-k axis <u>twice</u>. In Figure A.2a the locus of complex ω for real k is shown for both roots of ω from Equation A.8. The criteria of Reference 51 would certainly indicate that the "small loop" that encloses the saddle point of $\omega(k)$ represents an absolute instability because this contour (when considered alone) cannot be deformed into the upper-half ω-plane. This deformation in the ω-plane <u>can</u> be effected, however, if both contours are considered simultaneously. If we push the corresponding integration contour along the real axis in the k-plane to a line just above the point where $\partial\omega/\partial k = \infty$, as shown in Figure A.2b, the contours become as indicated. Both these contours now can be deformed into the upper-half ω-plane without encountering the saddle point of $\omega(k)$. This numerical ex-

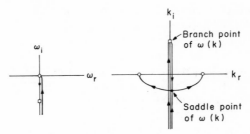

Branch point of $\omega(k)$ at $\omega = -j\frac{5}{4}$, $k = j\frac{3}{4}$
Saddle point of $\omega(k)$ at $\omega = -j\frac{1}{4}$, $k = -j\frac{1}{4}$

Figure A.1. Locus of roots in k-plane.

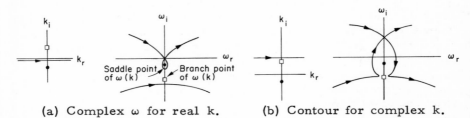

(a) Complex ω for real k.   (b) Contour for complex k.

Figure A.2. Contours of $\omega(k)$.

ample illustrates the importance of the branch points of $\omega(k)$ when the method of Reference 51 is employed.

Appendix B

THE LANDAU CONTOUR AND THE STABILITY CRITERIA
FOR A HOT PLASMA

This appendix is concerned with the dispersion equations of the type arising in the theory of waves in a hot, collisionless plasma. These dispersion equations have branch lines in the complex k-plane (for fixed complex ω) and therefore fall into a class that was excluded from consideration in the development of the general stability criteria of Chapter 2. Derfler[56] has considered many of the ideas presented here in his comprehensive treatment of the gap impedance in a warm plasma, and Kusse[72] has recently considered in detail the effect of the hot-plasma branch lines on the stability criteria of Chapter 2. Only a summary of the essential features of the problem are given in this appendix.

We consider explicitly the dispersion equation for longitudinal waves in a one-dimensional system:

$$\Delta(\omega, k) = 1 - \sum_n \omega_{pn}^2 \int_{-\infty}^{+\infty} \frac{f_{0n}(v_z)\, dv_z}{(\omega - kv_z)^2}$$

$$= 1 - \sum_n \frac{\omega_{pn}^2}{k^2} \int_{-\infty}^{+\infty} \frac{\frac{\partial f_{0n}}{\partial v_z}\, dv_z}{\left(v_z - \frac{\omega}{k}\right)} \tag{B.1}$$

(see Equations 3.3 and 3.4). The extension of this discussion to the other warm-plasma dispersion equations considered in Chapters 3 and 5 follows in a straightforward manner.

The dispersion equation (B.1) actually defines two functions of ω and k, depending on whether Im (ω/k) is assumed positive or negative when the velocity integration is carried out. We shall denote by $\Delta_1(\omega, k)$ the function obtained when Im (ω/k) is assumed to be negative, and by $\Delta_2(\omega, k)$ the function corresponding to Im (ω/k) assumed to be positive. As an example, for the simple resonance distribution given by Equation 3.8, we find

$$\Delta_1(\omega, k) = 1 - \sum_n \frac{\omega_{pn}^2}{(\omega - jkV_{Tn})^2} \tag{B.2}$$

143

and

$$\Delta_2(\omega, k) = 1 - \sum_n \frac{\omega_{pn}^2}{(\omega + jkV_{Tn})^2} \tag{B.3}$$

by simple residue techniques of evaluating the velocity integration.

The function $\Delta_1(\omega, k)$ can, of course, be analytically continued into the region Im $(\omega/k) > 0$; and, similarly, $\Delta_2(\omega, k)$ can be analytically continued for Im $\omega/k < 0$. For the resonance distribution, the algebraic expressions in Equations B.2 and B.3 make the analytic continuation a trivial matter. In general, the expression for the analytic continuation of $\Delta_1(\omega, k)$ as an integral over v_z in the region Im $(\omega/k) > 0$ is of the same form as Equation B.1, except that the velocity integration must extend out into the <u>complex</u> v_z-plane along the path indicated in Figure B.1.

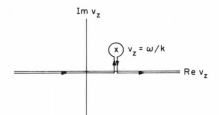

Figure B.1. Velocity integration for analytic continuation of $\Delta_1(\omega, k)$ into region Im $(\omega/k) > 0$.

The analytic continuation indicated in Figure B.1 is, of course, the same as that considered by Landau for the case of <u>real positive k</u>.[26,60] Landau, in his classical solution of the plasma oscillation problem, noted that for a fixed <u>positive</u> real k "imposed" on the system, the dispersion equation $\Delta_1(\omega, k)$ was the "proper one" in the context of an initial-value problem because Im (ω/k) was negative along the entire Laplace contour. In order to compute the asymptotic response of a <u>stable</u> plasma in a simple fashion, Landau showed that the important root of the dispersion relation was the least-damped root of $\Delta_1(\omega, k) = 0$ (again, for positive real k).

The stability criteria of Chapter 2 require an investigation of the roots of $\Delta(\omega, k) = 0$ in the complex k-plane for various complex values of ω with $\omega_i < 0$. In the present case, the dispersion equation (B.1) is discontinuous along the line Im $(\omega/k) = 0$; that is, it has a branch line in the k-plane along the line $(k_i/k_r) = (\omega_i/\omega_r)$, as shown in Figure B.2. Physically, this branch line arises because of the infinite number (continuum) of normal modes in a hot plasma, as was pointed out by Van Kampen[55] and very clearly by Dawson[73] from his multibeam approach.

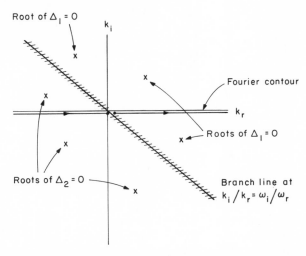

Figure B.2. Singularities of $1/\Delta$ in the k-plane for some $\omega = \omega_r + j\omega_i$ on the Laplace contour ($\omega_i < 0$).

We note in Figure B.2 that the Fourier contour for the computation of $F(\omega, z)$ (defined in Chapter 2) passes directly through
the branch line at $k = 0$. As was pointed out by Derfler,[56] this
circumstance is avoided if we do not allow any plasma particles
to travel faster than some speed c. He showed that a small
"gap" in the branch line around the $k = 0$ point is opened up when
$f_0(v_z) = 0$ for $v_z > c$. We can then formally let $c \to \infty$ and consider the Fourier integral as an integral from $-\infty$ to 0^- and from
0^+ to $+\infty$. We should, however, bear in mind that the $k = 0$ point
on the branch line really consists of two branch points located
very close together, and therefore this point on the branch line
cannot be "shifted around" by the techniques of analytic continuation to be discussed shortly.

If we close the Fourier integration in the lower-half k-plane
(for $z > d$), as was done in Chapter 2 in order to express $F(\omega, z)$
as a sum of normal modes (Figure 2.5), we find that the integration is now reduced to a sum of residues plus the contribution of the branch line (Figure B.3). Note also that the roots
of $\Delta(\omega, k) = 0$ which enter in the residue evaluation are those
(in the lower-half k-plane) which either have Im $(\omega/k) < 0$ and
are solutions of $\Delta_1 = 0$, or have Im $(\omega/k) > 0$ and are solutions
of $\Delta_2 = 0$. Any other roots are "spurious roots" [they are on
the wrong sheet of the function $\Delta(\omega, k)$[56]] and do not enter in
the residue evaluation of $F(\omega, z)$.

We can, however, "rotate" the branch line in the k-plane by
analytically continuing Δ_1 and Δ_2. This involves, in general, a
deformation of the velocity integration into the complex v_z-plane,
as was shown in Figure B.1. In this process, some of the poles
that were previously included in the residue evaluation of $F(\omega, z)$,

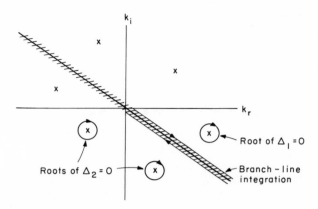

Figure B.3. Deformed Fourier contour for z > d and ω on the Laplace contour. (Integral is then expressed as a sum over the appropriate residues plus the integration along the branch line.)

as illustrated in Figure B.3, may now become "covered" and lie on the wrong sheet of $\Delta(\omega, k)$, and vice versa. That is, there is a certain lack of uniqueness in the division between the "discrete normal modes" contribution to $F(\omega, z)$ (poles of $1/\Delta$) and the branch-line contribution to $F(\omega, z)$, because this rotation of the branch line may result in a loss (or gain) of the pole contributions to $F(\omega, z)$,* as was pointed out by Derfler.[56]

On the other hand, there does <u>not</u> appear to be any real ambiguity concerning the roots that cross the real k-axis for ω values in the lower-half plane ("unstable wave" solutions). The branch line <u>cannot</u> be moved through the (deformed) Fourier contour C in the k-plane (see Figure 2.9). Therefore, a root such as A illustrated in Figure B.4 cannot be "covered up" by a rotation (or deformation) of the branch line unless we deform part of the branch line into the upper-half k-plane, where it makes a growing contribution to $F(\omega, z)$. It is <u>convenient</u> to keep the entire lower section of the branch line in the lower-half k-plane as shown in the figure, because the pole contributions then contain all the information regarding growing waves. Note also that a root that follows a trajectory like the one of root B in Figure B.4 <u>forces</u> a deformation of the branch line (an analytic continuation of the dispersion equation) because it "pinches" the C contour against the branch line. It is also clear from this discussion that the "collision" of roots through the

---

*The branch-line contribution to $F(\omega, z)$, of course, must also change when the branch line is rotated in such a manner that the total function $F(\omega, z)$ is the same, independent of the manner of computing it.

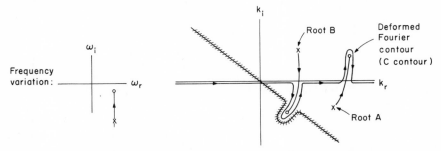

Note: Roots A and B are assumed to be "proper" ones; that is, they are solutions to $\Delta_1 = 0$ for ω on the Laplace contour.

Figure B.4. Locus of "unstable" roots of the dispersion equation.

C contour (signifying an absolute instability) has no ambiguity as to whether or not the colliding roots could have been "covered up" by a branch-line deformation.

In the general case, therefore, the application of the stability criteria to a hot plasma could be a very involved task, since, in applying the ideas of the previous paragraph, we must still keep track of which roots to the dispersion equation(s) are "proper ones" and which ones are "spurious" roots. In this monograph, however all of the examples in which the branch line is important are limited to the "weak-beam" case.* The unstable roots are then located very close to the real-k axis, and we can have recourse to the simple rule of using the $\Delta_1 = 0$ dispersion equation for $k_r > 0$ and the $\Delta_2 = 0$ dispersion equation for $k_r < 0$.

---

*Note that many of the hot-plasma dispersion equations that are analyzed, such as the square distribution or the expanded Maxwellian distribution in Section 5.2, have $\Delta_1 = \Delta_2$.

Appendix C

DERIVATION OF ONE-DIMENSIONAL DISPERSION EQUATION

In this appendix the dispersion equation for longitudinal and transverse waves along the steady magnetic field will be derived from an analysis of a particle orbit. In this approach each species of charged particles (electrons and ions) is considered to be made up of an infinite number of "cold beams," where each beam has a <u>unique</u> zero-order or unperturbed velocity and a constant (infinitesimal) unperturbed density in space.

The unperturbed trajectory of a particular particle, in the collisionless approximation, is a helix about the steady magnetic field. We identify the "nth beam" as that group of particles in which, at a particular instant of time, all particles have the <u>same</u> unperturbed velocity $\overline{v}_{0n}$ (both in magnitude <u>and</u> direction). The spatial distribution of these particles is constant, since the plasma is assumed to be uniform; that is, the unperturbed density ρ_{0n} is a constant. Note that the unperturbed velocity of the nth beam <u>is</u> a function of time in this approach, since the particles in that beam are spiraling about the steady magnetic field $\overline{B}_0$. Note also that the variables associated with the nth beam are, in a sense, "microscopic" variables; the collective, or "macroscopic" current and charge are obtained by summing over all such "microscopic" currents and charges. This point is important, because, as we shall see, even though the time variation of the "macroscopic" variables is sinusoidal at the frequency ω, the time variation of the "microscopic" variables is not necessarily at this same frequency, even within the linearized approximation.

The dispersion equation that we shall derive has been obtained by a number of authors from the collisionless Boltzmann-Vlasov equation (see Stix[26] and references therein). The derivation of this equation by use of the perturbed distribution function has been carried through both in Eulerian and Lagrangian variables. The method used here, although lacking in its generality and mathematical sophistication, perhaps has the advantage of presenting a somewhat clearer view of the physical phenomena involved, since the orbits of individual particles are considered explicitly.

The force equation for each particle is

$$\frac{d\overline{v}}{dt} = \frac{\partial \overline{v}}{\partial t} + (\overline{v} \cdot \nabla)\overline{v} = \frac{q}{m}(\overline{E} + \overline{v} \times \overline{B}) \tag{C.1}$$

where the macroscopic fields are used for $\overline{E}$ and $\overline{B}$ in the collisionless approximation. Therefore, the equation for the zero-order velocity $\overline{v}_{on}(t)$ is

$$\frac{d\overline{v}_{on}(t)}{dt} + \frac{q}{m}\,\overline{B}_0 \times \overline{v}_{on}(t) = 0 \tag{C.2}$$

since $\overline{B}_0$ is assumed to be uniform and there is no zero-order electric field.

The first-order velocity of a particular beam is determined from

$$\frac{\partial \overline{v}_n(z,t)}{\partial t} + v_{on_z}\left[\frac{\partial \overline{v}_n(z,t)}{\partial z}\right] + \frac{q}{m}\,\overline{B}_0 \times \overline{v}_n(z,t)$$

$$= \frac{q}{m}\,\overline{E}(z,t) + \overline{v}_{on}(t) \times \overline{B}(z,t) \tag{C.3}$$

where quantities without the 0 subscript denote first-order variables. Each beam must also satisfy its own conservation of charge equation

$$\nabla \cdot \overline{J}_n(z,t) = -\frac{\partial \rho_n(z,t)}{\partial t} \tag{C.4}$$

where

$$\overline{J}_n = \rho_n \overline{v}_{on} + \rho_{on}\overline{v}_n \tag{C.5}$$

is the first-order current contributed by this particular beam and ρ_n is the first-order charge density. The total macroscopic current is obtained by summing the contributions from all such beams; that is,

$$\overline{J}(z,t) = \sum_n \overline{J}_n(z,t) \tag{C.6}$$

where $\overline{J}$ is the macroscopic current density, and the sum will eventually be allowed to pass over into an integral.

Only the waves along the magnetic field will be considered, where all macroscopic variables have the dependence $\exp j(\omega t - kz)$ when $\overline{B}_0 = \overline{i}_z B_0$. Maxwell's equations can then be written in the form

$$k\overline{i}_z \times \overline{E} = \omega\mu_0\overline{H} \tag{C.7}$$

$$k\overline{i}_z \times \overline{H} = -\omega\epsilon_0\left(\overline{E} + \frac{\overline{J}}{j\omega\epsilon_0}\right) \tag{C.8}$$

where

$$\overline{E}(z, t) = \text{Re } \overline{\underline{E}}e^{j(\omega t - kz)} \tag{C.9}$$

and similarly for the other variables. Combining Equations C.7 and C.8, we have

$$j\omega\epsilon_0\underline{E}_z + \underline{J}_z = 0 \tag{C.10}$$

and

$$\left(\frac{k^2c^2}{\omega^2} - 1\right)\overline{E}_T = \frac{1}{j\omega\epsilon_0}\,\overline{J}_T \tag{C.11}$$

where $\overline{E}_T$ and $\overline{J}_T$ are the components transverse to the z-direction. It is intuitively clear, and will be proved later, that the transverse current can be written as a function of only the transverse component of the electric field. Therefore, when these dependencies are determined, Equations C.10 and C.11 give the dispersion equations for the longitudinal and the transverse waves, respectively.

C.1 Rotating Coordinates

For waves along $\overline{B}_0$, advantage is gained by describing all vectors by their circularly polarized components; for example,

$$\begin{bmatrix} v_R \\ v_L \\ v_z \end{bmatrix} = \begin{bmatrix} 1 & j & 0 \\ 1 & -j & 0 \\ 0 & 0 & 1 \end{bmatrix} \begin{bmatrix} v_x \\ v_y \\ v_z \end{bmatrix} \tag{C.12}$$

or

$$\overline{v}_r = \widetilde{R} \cdot \overline{v} \tag{C.13}$$

in matrix notation, where $\overline{v}_r$ symbolizes the set of rotating coordinates. Here v_R is the right-handed component, and v_L is the left-handed component. Two operations that are important for the mathematical manipulations in the remainder of this appendix are

$$\widetilde{R} \cdot (\overline{i}_z \times \overline{v}) = \widetilde{R} \cdot \begin{bmatrix} 0 & -1 & 0 \\ 1 & 0 & 0 \\ 0 & 0 & 0 \end{bmatrix} \cdot \widetilde{R}^{-1} \cdot \overline{v}_r = \begin{bmatrix} j & 0 \\ 0 & -j \end{bmatrix} \begin{bmatrix} v_R \\ v_L \end{bmatrix} \tag{C.14}$$

where $\widetilde{R}^{-1}$ is the inverse of the $\widetilde{R}$ matrix. The other operation is

$$\overline{A} \cdot \overline{B} = (\widetilde{R}^{-1} \cdot A_r) \cdot (\widetilde{R}^{-1} \cdot \overline{B}_r) = A_z B_z + \tfrac{1}{2}(A_R B_L + A_L B_R)$$

$$(C.15)$$

where $\overline{A}$ and $\overline{B}$ are any two vectors.

The use of the circularly polarized components greatly simplifies the mathematical manipulations which follow; however, in interpreting the resulting equations one should bear in mind that these quantities are not <u>physical</u> variables but are quantities derived from the physical variables by a <u>complex</u> transformation (Equation C.12). In this regard, note that one does <u>not</u> take the real part of such a circularly polarized component (like v_L, for instance) to obtain the "actual" physical variable, as <u>is</u> done in the usual complex notation for the sinusoidal steady state (Equation C.9). We will, in the description of the first-order electromagnetic field variables, also take the "physical" variable $\overline{v}$ in Equation C.12 to be a complex amplitude in the sinusoidal steady-state sense; in this case the circularly polarized component $\overline{v}_r$ will be written with a bar underneath.

C.2 Zero-Order Orbits

If we multiply Equation C.2 by the $\overline{\overline{R}}$ matrix and use Equation C.14, we obtain

$$\left.\begin{array}{c} \dfrac{dv_{0R}}{dt} + j\omega_c v_{0R} = 0 \\[2em] \dfrac{dv_{0L}}{dt} - j\omega_c v_{0L} = 0 \\[2em] \dfrac{dv_{0z}}{dt} = 0 \end{array}\right\} \qquad (C.16)$$

where $\omega_c = q\, B_0/m$ carries the sign of the charge and the n subscript will be suppressed for simplicity of notation. The solution of Equation C.16 gives

$$\left.\begin{array}{c} v_{0R}(t) = v_{T0} e^{-j(\omega_c t + \phi)} \\[2em] v_{0L}(t) = v_{T0} e^{j(\omega_c t + \phi)} \\[2em] v_{0z} = const \end{array}\right\} \qquad (C.17)$$

where v_{T_0} is the magnitude of the transverse velocity and ϕ specifies the initial direction of the transverse velocity. The solution for v_{0R} and v_{0L} must be as given in Equation C.17 in order that

$$\left.\begin{aligned} v_{0x}(t) &= \frac{1}{2}(v_{0R} + v_{0L}) = v_{T_0}\cos(\omega_c t + \phi) \\[2mm] v_{0y}(t) &= \frac{1}{2j}(v_{0R} - v_{0L}) = -v_{T_0}\sin(\omega_c t + \phi) \end{aligned}\right\} \tag{C.18}$$

satisfy Equation C.2.

C.3 First-Order Orbits

With use of Equation C.7 in Equation C.3, the first-order velocity must satisfy

$$\frac{d\overline{v}(z,t)}{dt} + \omega_c \overline{i}_z \times \overline{v}(z,t) = \frac{q}{m}\left(\underline{\overline{E}} + \frac{k}{\omega}(\overline{v}_0(t) \cdot \underline{\overline{E}})\overline{i}_z - \frac{k}{\omega}v_{0z}\underline{\overline{E}}\right)e^{j(\omega t - kz)} \tag{C.19}$$

Multiplying Equation C.19 by $\widetilde{R}$, and using Equations C.14 and C.15, we have

$$\frac{dv_z}{dt} = \frac{q}{m}\left[\underline{E}_z + \frac{k}{2\omega}\left(v_{0R}(t)\underline{E}_L + v_{0L}(t)\underline{E}_R\right)\right]e^{j(\omega t - kz)} \tag{C.20}$$

$$\frac{dv_R}{dt} + j\omega_c v_R = \frac{q}{m}\left(1 - \frac{kv_{0z}}{\omega}\right)\underline{E}_R e^{j(\omega t - kz)} \tag{C.21}$$

$$\frac{dv_L}{dt} - j\omega_c v_L = \frac{q}{m}\left(1 - \frac{kv_{0z}}{\omega}\right)\underline{E}_L e^{j(\omega t - kz)} \tag{C.22}$$

Because of the linearity, we can compute separately the current arising from E_z, E_L, and E_R.

C.4 Longitudinal Waves

For only E_z nonzero, the first-order velocities are

$$\left.\begin{aligned} v_R &= v_L = 0 \\[4mm] v_z &= \frac{\dfrac{q}{m}}{j(\omega - kv_{0z})}\,\underline{E}_z e^{j(\omega t - kz)} \end{aligned}\right\} \tag{C.23}$$

The conservation of charge, Equation C.4, requires that

$$\frac{\partial \rho}{\partial t} + v_{0z} \frac{\partial \rho}{\partial z} = -\rho_0 \frac{\partial v_z}{\partial z} \tag{C.24}$$

from which we obtain

$$\rho(z, t) = \frac{k\rho_0}{\omega - kv_{0z}} v_z(z, t) \tag{C.25}$$

From Equations C.23 and C.25, the longitudinal current contributed by this particular beam is

$$\frac{J_{zn}}{j\omega\epsilon_0} = - \frac{\frac{q\rho_0}{\epsilon_0 m}}{(\omega - kv_{0z})^2} E_z \tag{C.26}$$

We now add the contributions from all such beams, noting that $\rho_0 \to \rho_{0T} f_0(\overline{v}_0)\, d^3\overline{v}_0$, where ρ_{0T} is the total zero-order charge density of that species and $f_0(\overline{v}_0)$ is the distribution of zero-order velocities. Note that although each individual beam <u>does</u> carry a <u>transverse</u> current of

$$J_{L_n} = \rho v_{T0} e^{j(\omega t - kz + \omega_c t + \phi)} \left.\rule{0pt}{60pt}\right\}$$

$$J_{R_n} = \rho v_{T0} e^{j(\omega t - kz - \omega_c t - \phi)} \tag{C.27}$$

the integration over all such beams gives no net <u>transverse</u> current if the distribution function of zero-order velocities is independent of ϕ [which is another way of saying that it is a function of only $(v_x^2 + v_y^2)^{1/2}$ and v_z]. Therefore, the combination of Equations C.26 and C.10 results in the well-known dispersion equation for longitudinal waves:

$$1 - \sum \omega_P^2 \int \frac{f_0(\overline{v}_0)\, d^3\overline{v}_0}{(\omega - kv_{0z})^2} = 0 \tag{C.28}$$

where $\omega_P^2 = q\rho_{0T}/\epsilon_0 m$ and the sum is over all species of charged particles (electrons and ions). The integration over the transverse velocities can be carried out directly to give

$$1 - \sum \omega_p^2 \int_{-\infty}^{+\infty} \frac{f_0(v_z)\ dv_z}{(\omega - kv_z)^2} \tag{C.29}$$

C.5 Transverse Waves

If we assume that only E_L is nonzero, then Equations C.20 and C.22 give the velocities as

$$\left.\begin{aligned}
\underline{v}_L &= \frac{\dfrac{q}{m}\left(1 - \dfrac{kv_{0z}}{\omega}\right)}{j(\omega - kv_{0z} - \omega_c)}\ \underline{E}_L e^{j(\omega t - kz)} \\[3mm]
\underline{v}_z &= \frac{\dfrac{1}{2}\dfrac{q}{m}\dfrac{k}{\omega}v_{T0}}{j(\omega - kv_{0z} - \omega_c)}\ \underline{E}_L e^{j[\omega - \omega_c)t - kz - \phi]} \\[3mm]
\underline{v}_R &= 0
\end{aligned}\right\} \tag{C.30}$$

It is interesting to note that purely transverse fields cause a longitudinal first-order velocity in the beam that oscillates at the frequency $(\omega - \omega_c)$. This longitudinal velocity arises from the $\overline{v}_{0T} \times \overline{B}$ force arising from the unperturbed velocity about the field lines. The conservation of charge equation (C.24) gives

$$\rho(z, t) = \frac{k\rho_0}{\omega - \omega_c - kv_{0z}}\ v_z(z, t) \tag{C.31}$$

From Equations C.30 and C.31, the left-polarized current arisin from a single beam can be derived in the form

$$J_L = \rho_0 v_L + \rho v_{L0}$$

$$= -j\omega\epsilon_0 E_L \frac{q\rho_0}{\epsilon_0 m\omega^2}\left[\frac{\omega - kv_{0z}}{\omega - \omega_c - kv_{0z}} + \frac{\dfrac{1}{2}k^2 v_{T0}^2}{(\omega - \omega_c - kv_{0z})^2}\right] \tag{C.32}$$

As it is easily shown that J_R and J_z are identically zero when integrated over all such beams, their explicit expression for a single beam will not be given. Therefore, the right- and left-polarized waves are uncoupled from each other as well as from the longitudinal waves. From Equations C.32 and C.11, we obtain the dispersion equation as

$$\frac{k^2c^2}{\omega^2} = 1 - \sum \frac{\omega_p^2}{\omega^2} \int \left[\frac{\omega - kv_{0z}}{\omega - \omega_c - kv_{0z}} + \frac{\frac{1}{2}k^2 v_{T0}^2}{(\omega - \omega_c - kv_{0z})^2} \right] f_0(\overline{v}_0)\, d^3\overline{v}_0$$

$$\text{(C.33)}$$

where the sum is over all species. Equation C.33 is for the left-polarized wave; for the right-polarized wave, let $\omega_c \rightarrow -\omega_c$. Using the fact that $f_0 = f_0(v_z, v_T^2)$, and omitting the zero subscript on the velocity, we can write Equation C.33 in the form

$$\frac{k^2c^2}{\omega^2} = 1 + \sum \frac{\omega_p^2}{\omega^2} \int_{-\infty}^{+\infty} \int_0^{\infty} \left[\frac{(\omega - kv_z)\frac{\partial f_0}{\partial v_T} + kv_T \frac{\partial f_0}{\partial v_z}}{\omega - \omega_c - kv_z} \right] \pi v_T^2\, dv_T\, dv_z$$

$$\text{(C.34)}$$

by performing an integration by parts. Equation C.34 can be simplified further for those species having a symmetrical distribution function, that is, for the case when $f_0 = f_0(v)$, where $v^2 = v_T^2 + v_z^2$. The term in the sum then becomes

$$-\frac{\omega_p^2}{\omega^2} \int \frac{f_0(v)}{\omega - kv_z - \omega_c}\, d^3\overline{v} \qquad \text{(C.35)}$$

The explicit expression for the dispersion equation of left-polarized waves in a beam-plasma system with a cold beam and for plasma particles with symmetrical distribution functions is

$$\frac{k^2c^2}{\omega^2} = 1 - \frac{\omega_{pe}^2}{\omega} \int_{-\infty}^{+\infty} \frac{f_{0e}(v_z)\, dv_z}{\omega - kv_z + \omega_{ce}} - \frac{\omega_{pi}^2}{\omega} \int_{-\infty}^{+\infty} \frac{f_{0i}(v_z)\, dv_z}{\omega - kv_z - \omega_{ci}}$$

$$- \frac{\omega_{pb}^2(\omega - kv_0)}{\omega^2(\omega - kv_0 + \omega_{ce})} \qquad \text{(C.36)}$$

where

$$f_{0e,i}(v_z) = \int_0^{\infty} f_{0e,i}(v) 2\pi v_T\, dv_T \qquad \text{(C.37)}$$

Appendix D

CLASSIFICATION OF LONGITUDINAL
WEAK-BEAM INSTABILITIES

If we are not close to a point of synchronism of a plasma wave with the beam, that is, if $K_{\parallel}(\omega = kv_0) \neq 0$, then the solution for the beam waves is of the form

$$k = \frac{\omega}{v_0} \pm \delta_k \tag{D.1}$$

where δ_k is, in general, complex and $\delta_k \to 0$ as $\omega_{pb} \to 0$. Since both roots in Equation D.1 are in the lower-half k-plane when ω has a negative imaginary part larger than the order of Im $(\delta_k v_0)$, it follows that there are no absolute instabilities in this range of ω, and that any solution with $k_i > 0$ is an amplifying wave.

If the frequency is very close to a point of synchronism, that is, if $K_{\parallel}(\omega = kv_0) = 0$ for some $\omega = \omega_0$ and $k = k_0 = \omega_0/v_0$, then the dispersion equation becomes

$$\frac{\beta_{pb}^2}{\left(k - \frac{\omega}{v_0}\right)^2} \simeq \left(\frac{\partial K_{\parallel}}{\partial k}\right)_0 \left[(k - k_0) - \frac{1}{v_{g0}}(\omega - \omega_0)\right] \tag{D.2}$$

where

$$v_{g0} = \left(\frac{\partial \omega}{\partial k}\right)_0 = -\frac{\left(\frac{\partial K_{\parallel}}{\partial k}\right)_0}{\left(\frac{\partial K_{\parallel}}{\partial \omega}\right)_0} \tag{D.3}$$

is the group velocity of the unperturbed plasma wave. If we let $\Gamma = k - k_0$ and $\Omega = \omega - \omega_0$, Equation D.2 can be written as

$$\left(\Gamma - \frac{\Omega}{v_0}\right)^2 \left(\Gamma - \frac{\Omega}{v_{g0}}\right) = \frac{\beta_{pb}^2}{\left(\frac{\partial K_{\parallel}}{\partial \omega}\right)_0} \tag{D.4}$$

Since for the plasma waves $v_{g0} > 0$ for all $k_0 > 0$, all <u>three</u> roots of k from Equation C.4 are in the lower-half k-plane for sufficiently large negative imaginary part of ω (or Ω). Again, it follows that there cannot be an absolute instability, and any root with $k_i > 0$ is an amplifying wave. This result is physically reasonable because the one-dimensional plasma waves are of the forward-traveling type near the synchronism with the beam, and hence, there is no "feedback mechanism" to generate an absolute instability.

Appendix E

AN INSTABILITY CONDITION FOR LOSSLESS SYSTEMS

We consider the dispersion equation for a lossless system

$$\Delta(\omega, k) = 0 \tag{E.1}$$

in which the complex roots of ω for real values of k must occur in complex conjugate pairs. An example might be an algebraic equation in ω and k with real coefficients, as considered in Section 3.4. We assume that the complex roots of ω for real k have their <u>onset</u> at some finite value of k, say $k = k_0$ and $\omega = \omega_0 \pm j\nu$ with $\nu \to 0$ as $k \to k_0$ (Figure E.1). Expanding Equation E.1 about the point (ω_0, k_0), we obtain

$$\left(\frac{\partial \Delta}{\partial \omega}\right)_0 (\omega - \omega_0) + \frac{1}{2}\left(\frac{\partial^2 \Delta}{\partial \omega^2}\right)_0 (\omega - \omega_0)^2 + \left(\frac{\partial \Delta}{\partial k}\right)_0 (k - k_0) + \cdots \simeq 0 \tag{E.2}$$

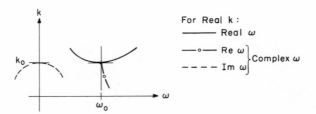

Figure E.1. Illustration of onset of complex ω in a lossless system.

In order that the point (ω_0, k_0) be a <u>double root</u> of ω for $k = k_0$, a necessary condition if it is to be the onset of a pair of complex conjugate roots, we must have

$$\left(\frac{\partial \Delta}{\partial \omega}\right)_0 = 0 \tag{E.3}$$

Therefore, the condition for an onset of unstable modes in a lossless system is that there exist some <u>real</u> ω and k which satisfy the original dispersion equation $(\overline{E.1})$ <u>and</u> the condition given by Equation E.3.

Since

$$\frac{\partial \Delta}{\partial \omega} = - \frac{\partial \Delta}{\partial k} \frac{\partial k}{\partial \omega} \qquad (E.4)$$

the instability condition (E.3) is equivalent to the statement that there must be a point of "infinite group velocity" ($\partial \omega / \partial k = \infty$) of one of the propagating waves, as is clear from the sketch in Figure E.1.

Appendix F

TRANSVERSE BEAM WAVES

The solution of Equation 3.77 is complicated because of its cubic nature. The propagating modes are most easily obtained by transforming into a reference frame that moves with the beam. In this reference frame an observer sees a stationary electron plasma, which has a dispersion equation quadratic in the wave number k. The waves in the laboratory frame are easily obtained by a relativistic transformation of (ω, k) as

$$
\left.
\begin{aligned}
\omega &= \gamma_0(\omega' + k'v_0) \\[2mm]
k &= \gamma_0\left(k' + \omega' \frac{v_0}{c^2}\right)
\end{aligned}
\right\}
\tag{F.1}
$$

where (ω', k') refers to the moving frame and (ω, k) refers to the laboratory frame. The dispersion equation in the moving frame is

$$
k'^2 = \frac{\omega'^2}{c^2}\left[1 - \frac{\omega_{pb}^2}{\omega'(\omega' \pm \omega'_{ce})}\right]
\tag{F.2}
$$

where ω'_{ce} is computed using the rest mass and ω_{pb} is invariant under the transformation because n_b/m_t is invariant.

The dispersion equation for the left-polarized wave from Equation F.2 is sketched in Figure F.1 for a relatively weak beam $(\omega_{pb} \ll \omega_{ce})$. The dispersion of the right-polarized wave is obtained by letting $\omega' \to -\omega'$. The transformation to the laboratory frame gives the dispersion curve shown in Figure 3.14. We note that the left-polarized wave has a very low frequency backward-wave branch with phase velocity less than v_0 under these conditions; it is shown in more detail in Figure 3.15.

Since, as will be argued later, the electron beam is a passive medium when viewed from the moving (primed) reference frame, we can distinguish the active and passive waves by inspection of the dispersion diagram, following a method originated by Sturrock. The relativistic transformation of small-signal energy between the two reference frames is given by[74-76]

160

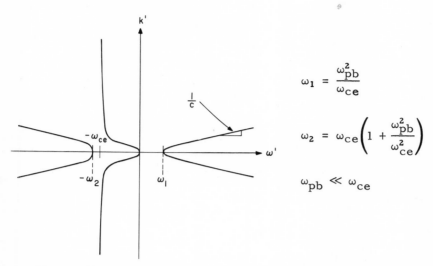

$$\omega_1 = \frac{\omega_{pb}^2}{\omega_{ce}}$$

$$\omega_2 = \omega_{ce}\left(1 + \frac{\omega_{pb}^2}{\omega_{ce}^2}\right)$$

$$\omega_{pb} \ll \omega_{ce}$$

Figure F.1. Dispersion of left-polarized wave in moving reference frame.

$$\frac{W}{\omega}\left(1 - \frac{v_g^2}{c^2}\right)^{\frac{1}{2}} = \frac{W'}{\omega'}\left(1 - \frac{v_g'^2}{c^2}\right)^{\frac{1}{2}} \qquad (F.3)$$

where W is the small-signal energy (per unit volume) and $v_g = \partial\omega/\partial k$ is the group velocity of the wave in that reference frame. In the primed reference frame, an observer sees a stationary electron plasma, which physically must be passive (that is, it must have $W' \geq 0$). Therefore, an active wave in the laboratory frame (W < 0) is obtained only when

$$\frac{\omega'}{\omega} < 0 \qquad (F.4)$$

or [74]

$$0 < \frac{\omega}{k} < v_0 \qquad (F.5)$$

From Figure 3.14 we see that only the left-polarized waves can be active. There are two active wave branches: the low-frequency region with small wave number k, and the "slow cyclotron wave" that has $k \simeq (\omega + \omega_{ce})/v_0$.

Appendix G

QUASI-STATIC DISPERSION EQUATION

The dispersion equation for quasi-static waves in a beam-plasma system will be derived in this appendix. The determinantal equation for the unfilled circular waveguide will also be given. These results have been derived previously by several authors;[1,29] in the present derivation the advantage of formulating the equations in terms of the charge density rather than the current density will be stressed.

We consider first a single species of charged particles that move with unperturbed velocity v_n parallel to a magnetic field $\overline{B}_0$ that is aligned with the axis z of a waveguide structure (Figure 4.1). This species is characterized by the plasma and cyclotron frequencies ω_{pn} and ω_{cn}, where ω_{cn} carries the sign of the charge. The Doppler shifted frequency in a reference frame which moves with the velocity v_n (the rest frame of that species) is

$$\omega_n = \omega - \beta v_n \tag{G.1}$$

in the nonrelativistic (slow-wave) approximation.

The current density of this species in its rest frame is[61*]

$$\overline{J}_n = j\omega_n \overline{P}_n \tag{G.2}$$

where

$$\overline{P}_n \equiv \epsilon_0 (\chi_{\perp n} \overline{E}_T + \chi_{xn} \overline{i}_z \times \overline{E}_T + \chi_{\parallel n} E_z \overline{i}_z) \tag{G.3}$$

and where $\overline{E}_T$ and E_z are the transverse and longitudinal components of the electric field. In Equation G.3,[61]

---

*The vector $\overline{P}$ as defined here can be interpreted as the usual polarization vector in the case of a polarizable medium, and, in fact, all of the relations between $\overline{P}$ and ρ given here follow directly from the well-known electromagnetic theory of material media.[77]

$$\chi_{\perp n} = -\frac{\omega_{pn}^2}{\omega_n^2 - \omega_{cn}^2}$$

$$\chi_{xn} = -\frac{j\omega_{pn}^2 \omega_{cn}}{\omega_n(\omega_n^2 - \omega_{cn}^2)} \left.\right\} \qquad \text{(G.4)}$$

$$\chi_{\|n} = -\frac{\omega_{pn}^2}{\omega_n^2}$$

From the conservation of charge, the charge density introduced by this species of particles is

$$\rho_n = -\nabla \cdot \overline{P}_n \qquad\qquad \text{(G.5)}$$

In a nonrelativistic transformation of coordinates, ρ_n is invariant. This is not true of the current density $\overline{J}_n$. This is the reason why it is advantageous to deal with charge density in a quasi-static formulation because only the charge density need be determined in order to find the scalar potential φ, where $\overline{E} = -\nabla\varphi$. In fact, the application of Gauss's law now yields

$$\epsilon_0 \nabla^2 \varphi = -\sum_n \rho_n \qquad\qquad \text{(G.6)}$$

or

$$\nabla^2 \varphi = -\sum_n (\chi_{\perp n} \nabla_T^2 \varphi - \beta^2 \chi_{\|n} \varphi) \qquad\qquad \text{(G.7)}$$

If we write out this sum explicitly for a cold-beam-plasma system, the dispersion equation (4.2) is obtained for the filled waveguide.

In the case of an unfilled waveguide, there is a surface charge of

$$\sigma_n = \overline{n} \cdot (\overline{P}_{n1} - \overline{P}_{n2}) \qquad\qquad \text{(G.8)}$$

on any boundary where $\overline{P}_n$ is discontinuous. Once again, this surface charge is invariant in a nonrelativistic transformation. If we define

$$\overline{D} = \epsilon_0 \overline{E} + \sum_n \overline{P}_n \qquad (G.9)$$

it is easily shown from Equation G.8 and the conservation of charge that the normal component of $\overline{D}$ must be continuous across any boundary. In applying this boundary condition, one should note that only the species that are present in a given region enter in the sum in Equation G.9.

In the case of a circular system as shown in Figure 4.5, the potentials are

$$\varphi_i = A \frac{J_n(pr)}{J_n(pb)} e^{jn\phi}, \qquad r < b \qquad (G.10)$$

$$\varphi_0 = A \frac{N_n(qr) - \dfrac{N_n(qa)}{J_n(qa)} J_n(qr)}{N_n(qb) - \dfrac{N_n(qa)}{J_n(qa)} J_n(qb)} e^{jn\phi}, \qquad r > b \qquad (G.11)$$

where p and q are defined in Sections 4.1 and 4.2. The boundary condition on $\overline{D}$ requires that

$$K_{\perp}^0 \frac{\partial \varphi_i}{\partial r} (r = b) + \frac{n}{b} \frac{\omega_{pb}^2 \omega_{ce}}{\omega_d(\omega_d^2 - \omega_{ce}^2)} \varphi_i(r = b) = K_{\perp} \frac{\partial \varphi_0}{\partial r} (r = b) \qquad (G.12)$$

where $\omega_d = \omega - \beta v_0$ and the other symbols are defined in Section 4.1. Therefore, we obtain finally[29]

$$K_{\perp}^0 pb \frac{J_n'(pb)}{J_n(pb)} + n \frac{\omega_{pb}^2 \omega_{ce}^2}{\omega_d(\omega_d^2 - \omega_{ce}^2)} = K_{\perp} qb \frac{N_n'(qb) - \dfrac{N_n(qa)}{J_n(qa)} J_n'(qb)}{N_n(qb) - \dfrac{N_n(qa)}{J_n(qa)} J_n(qb)}$$

$$(G.13)$$

where the prime on a Bessel function indicates the first derivative with respect to its argument.

Appendix H

ABSOLUTE INSTABILITY OF SPACE-CHARGE WAVES

The dispersion equation near the coupling of the space-charge waves with the backward plasma wave ($\omega \simeq \omega_2$) is of the form

$$\left(\beta - \frac{\omega}{v_0}\right)^2 [\beta - \beta_0(\omega)] = \beta_e^3 C^3 \tag{H.1}$$

in the weak beam (that is, small C) approximation. We now ex-. pand this dispersion equation near the intersection frequency ω_2, where $\beta_0(\omega_2) = \omega_2/v_0$. If we let

$$v_p = -\left[\frac{\partial \beta_0(\omega)}{\partial \omega}\right]^{-1}_{\omega=\omega_2} \tag{H.2}$$

and introduce the normalized variables

$$\Gamma = \frac{\beta - \dfrac{\omega_2}{v_0}}{\beta_e C} \tag{H.3}$$

$$\Omega = \frac{\omega - \omega_2}{\beta_e C v_p} \tag{H.4}$$

and

$$\alpha = \frac{v_p}{v_0} \tag{H.5}$$

then the dispersion equation becomes

$$(\Gamma - \alpha\Omega)^2(\Gamma + \Omega) = 1 \tag{H.6}$$

The velocity v_p is the group velocity of the plasma wave near the intersection and is positive according to its definition in Equation H.2. We first show that there is a saddle point of $\Omega(\Gamma)$ in the lower-half Ω-plane. In order that $\partial\Omega/\partial\Gamma = 0$, we must have

$$\Gamma = -\left(\frac{2 - \alpha}{3}\right)\Omega \qquad (H.7)$$

From the dispersion equation (H.6), we find that the saddle point of $\Omega(\Gamma)$ of interest is at

$$\Omega_s = \frac{3}{4^{\frac{1}{3}}(1 + \alpha)}\left(-\frac{1}{2} - j\frac{\sqrt{3}}{2}\right) = \frac{3}{4^{\frac{1}{3}}(1 + \alpha)}\,e^{-j2\pi/3} \qquad (H.8)$$

in the Ω-plane.

To apply the criterion on absolute instabilities, it is convenient in this case to solve for the roots of Γ as Ω varies along a line at $-120°$ to the positive real-Ω axis (Figure H.1). Therefore, we let

$$\Omega = We^{-j2\pi/3} \qquad (H.9)$$

where W is considered to be a real number. We attempt a solution for Γ in the form

$$\Gamma = ge^{-j2\pi/3} \qquad (H.10)$$

with g considered as real. The relation between g and W can be written in the form

$$\left.\begin{aligned} \frac{1}{(g - \alpha W)^2} &= g + W \\[2mm] f_L(g) &= f_R(g) \end{aligned}\right\} \qquad (H.11)$$

Both sides of Equation H.11 are plotted in Figure H.2. For large positive W, there are three real roots for g. As W decreases, it clear from the figure that a critical value of W will be reached

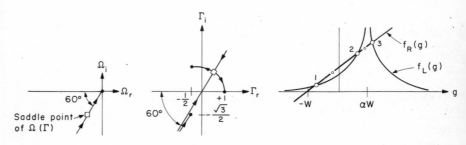

Figure H.1. Locus of roots of Γ.   Figure H.2. Plot of Equation H.11

where the two roots labeled 1 and 2 merge. This frequency is exactly that required for a double root of Γ (or g); that is, it is the frequency given by Equation H.8. For $\Omega = 0$, the three roots of Γ are

$$\Gamma = \begin{cases} +1 \\ -\dfrac{1}{2} \pm j\,\dfrac{\sqrt{3}}{2} \end{cases} \tag{H.12}$$

With this information, the loci of the roots of Γ can be sketched as shown in Figure H.1, where we have illustrated the case $\alpha < 2$, for which the double root is in the upper-half Γ-plane (see Equation H.7). We see that the merging roots of Γ come from opposite halves of the Γ-plane when Ω has a large negative imaginary part, and hence it does correspond to an absolute instability. It is interesting to note also that the merging roots are, in a sense, the "slow space-charge beam wave" and "backward plasma wave," as would be expected physically. (This identification of roots is made, as usual, for large $|\Omega|$.)

This example illustrates a point that was discussed in Section 2.6 concerning the meaning of "amplifying waves" when an absolute instability is present. For frequencies well below ω_2, there is amplification of the slow space-charge wave owing to the reactive-medium effect. However, from Figure H.1, it is clear that there must be an abrupt switch of the root with $\beta_i > 0$ for real ω from an "amplifying" to an "evanescent" wave. The exact real frequency at which this occurs depends on the manner of analytic continuation of $\bar{F}(\omega, z)$ to the real ω-axis, as discussed in Section 2.6.4. (In Figure H.1, where $\Delta\Omega_i/\Delta\Omega_r = \sqrt{3}$ during the analytic continuation, it occurs at $\Omega_r = 0$.) If we consider a beam-plasma system that is shorter than the minimum starting length for oscillation, then it is meaningful to speak of amplification at real frequencies. The significance of the roots of complex β for real ω near the synchronous frequency ω_2 is not clear within the context of the criteria of Chapter 2. Nevertheless, the presence of this "abrupt transition" does cast some doubt on the validity of equating the curve of β_i vs. real ω with a theoretical gain curve for a beam-plasma amplifier.[30-32]

It might also be mentioned that this case of three-wave coupling is a counterexample to the Sturrock criterion, which states that "if complex β for real ω is obtained from the dispersion equation, then the instability is convective."[46]

Finally, a similar analysis of the dispersion equation near $\omega = \omega_1$ shows that no absolute instability is obtained, and that the root with $\beta_i > 0$ for real ω is an amplifying wave (in the weak-beam approximation). This result is almost immediately obvious; since $\partial\beta_0/\partial\omega > 0$ near ω_1, all three roots of β will be in the lower-half β-plane for complex ω with a large enough negative imaginary part, and therefore the roots cannot possibly merge through the C contour.

Appendix I

MONOTONICALLY DECREASING CHARACTER OF THE
p^2 VS. q^2 RELATION

We consider an electron beam that is homogeneous over its cross section but of an arbitrary shape (Figure I.1). The plasma

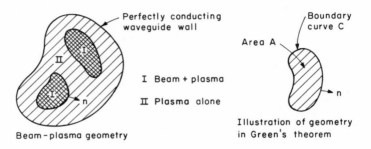

Figure I.1. Cross-sectional geometries.

is homogeneous and fills the waveguide structure. Inside the beam regions, we have

$$\nabla_T^2 \varphi_i + p^2 \varphi_i = 0 \tag{I.1}$$

while outside the beam regions,

$$\nabla_T^2 \varphi_0 + q^2 \varphi_0 = 0 \tag{I.2}$$

We shall consider p^2 and q^2 to be <u>real</u> in this proof. On the boundaries between any two regions, we have the continuity conditions

$$\varphi_i = \varphi_0 \tag{I.3}$$

and

$$\bar{n} \cdot \nabla_T \varphi_i = \bar{n} \cdot \nabla_T \varphi_0 \tag{I.4}$$

where $\bar{n}$ is a unit normal to the boundary. Green's theorem in two dimensions states that

$$\int_A \nabla_T \cdot \overline{D} \, da = \oint_C \overline{n} \cdot \overline{D} \, ds \qquad (I.5)$$

where $\overline{D}$ is any vector and $\overline{n}$ is now a unit normal pointing out of the area A bounded by the curve C (Figure I.1). If we let $\overline{D} = g\nabla_T f$, where g and f are any two scalar functions, we have

$$\int_A \nabla_T g \cdot \nabla_T f \, da + \int_A g\nabla_T^2 f \, da = \oint_C g\overline{n} \cdot \nabla_T f \, ds \qquad (I.6)$$

By interchanging g and f (that is, letting $\overline{D} = f\nabla_T g$) and subtracting the two equations, we obtain the second Green's theorem

$$\int_A (g\nabla_T^2 f - f\nabla_T^2 g) \, da = \oint_C (g\overline{n} \cdot \nabla_T f - f\overline{n} \cdot \nabla_T g) \, ds$$
$$(I.7)$$

We now consider small first-order variations in p^2, q^2, φ_i, and φ_0. Taking a first-order variation of Equation I.1, we obtain

$$\nabla_T^2(\delta\varphi) + \delta(p^2)\varphi + p^2(\delta\varphi) = 0 \qquad (I.8)$$

either inside or outside the beam regions (with $p^2 \to q^2$ outside). If we let $g = \delta\varphi^*$ and $f = \varphi$ in Equation I.7, we find, using Equations I.8 and I.1, that

$$\delta(p^2) \int_A |\varphi|^2 \, da = \oint_C \overline{n} \cdot [(\delta\varphi^*)\nabla_T\varphi - \varphi\nabla_T(\delta\varphi^*)] \, ds$$
$$(I.9)$$

since we are assuming that both p^2 and $\delta(p^2)$ are real. If we now write similar equations for all regions both inside and outside the beam and add all of these equations, the integrations around the curves C joining the regions cancel because of the continuity conditions (Equations I.3 and I.4 and the first-order variations of these equations). (Note that $\overline{n}$ points in opposite directions for two adjoining regions.) Therefore, we have the result that

$$\delta(p^2) \int_{A_b} |\varphi_i|^2 \, da + \delta(q^2) \int_{A_0} |\varphi_0|^2 \, da = 0 \qquad (I.10)$$

where A_b is the total beam area I and A_0 is the total area of the regions outside of the beam II. Equation I.10 can be written in the form

$$\frac{\partial(p^2)}{\partial(q^2)} = - \frac{\displaystyle\int_{A_0} |\varphi_0|^2 \, da}{\displaystyle\int_{A_b} |\varphi_i|^2 \, da} \leq 0 \qquad (I.11)$$

Therefore, the curves of p^2 vs. q^2 derived from the boundary conditions must always be monotonically decreasing.

Appendix J

DISPERSION EQUATION FOR RESISTIVE-
MEDIUM AMPLIFICATION

If we assume only a longitudinal temperature for the plasma electrons, and assume that the beam electrons and plasma ions are cold, the extension of Equation 5.1 takes the form

$$p^2 = -\beta^2 \frac{K_\parallel^0}{K_\perp^0} \tag{J.1}$$

where

$$K_\perp^0 = K_{\perp i} - \omega_{pe}^2 \int_{-\infty}^{+\infty} \frac{f_{0e}(v_z) \, dv}{(\omega - \beta v_z)^2 - \omega_{ce}^2} \tag{J.2}$$

and the other symbols are defined in Equations 5.2, 5.3, and 5.4. In Equation J.2 we have neglected the transverse motion of the beam electrons because we are interested in the space-charge wave interaction of a very weak beam. For a resonance distribution function of the form (see Equation 3.8),

$$f_{0e}(v_z) = \frac{V_{Te}}{\pi} \left(\frac{1}{v_z^2 + V_{Te}^2} \right) \tag{J.3}$$

this dispersion equation becomes

$$p^2 \left[K_{\perp i} - \frac{\omega_{pe}^2}{(\omega - j\beta V_{Te})^2 - \omega_{ce}^2} \right] = -\beta^2 \left[K_{\parallel i} - \frac{\omega_{pb}^2}{(\omega - \beta v_0)^2} - \frac{\omega_{pe}^2}{(\omega - j\beta V_{Te})^2} \right] \tag{J.4}$$

for the $\beta_r > 0$ region (see Appendix B). This result is obtained most easily by closing the integration over v_z in the upper-half complex v_z-plane, thus reducing the integration to the evaluation of the residue at the simple pole at $v_z = jV_{Te}$. For a very weak beam, we can solve for the space-charge waves by setting $\beta = \omega/v_0$ everywhere except in the beam term, and we then obtain Equation 5.13.

Appendix K

CONDITION FOR AN ABSOLUTE INSTABILITY

In this appendix the condition for a double root of complex β for real ω will be formulated. The dispersion equation for the hot-electron plasma is

$$\Delta(\omega, \beta) = 1 - \frac{\omega_{pi}^2}{\omega^2} + \frac{\beta_{De}^2 + p^2 K_{\perp i}}{\beta^2} - \frac{\beta_{pb}^2}{\left(\beta - \dfrac{\omega}{v_0}\right)^2} = 0 \quad (K.1)$$

The condition for a double root of β is that $\partial\Delta/\partial\beta = 0$. This requires that

$$\beta_{De}^2 + p^2 K_{\perp i} = \frac{\beta^3 \beta_{pb}^2}{\left(\beta - \dfrac{\omega}{v_0}\right)^3} \quad (K.2)$$

Equations K.1 and K.2 must be satisfied simultaneously. When we impose the additional constraint that ω be real and β be complex, we shall find that only certain values of the parameters are possible for this situation to prevail. In other words, the system parameters must take on certain critical values in order that the saddle point of $\omega(\beta)$ that is of interest lie on the real-ω axis.

By using Equation K.2 in Equation K.1, we obtain

$$\left(\beta - \frac{\omega}{v_0}\right)^3 = \left(\frac{\omega}{v_0}\right)^3 \left(\frac{\dfrac{\omega_{pb}^2}{\omega_{pi}^2}}{1 - \dfrac{\omega^2}{\omega_{pi}^2}}\right) \quad (K.3)$$

If we let

$$x = \left(\frac{\dfrac{\omega_{pb}^2}{\omega_{pi}^2}}{1 - \dfrac{\omega^2}{\omega_{pi}^2}}\right)^{\frac{1}{3}} \quad (K.4)$$

172

the <u>complex</u> roots of Equation K.3 are

$$\beta = \frac{\omega}{v_0} (1 + xe^{\pm j2\pi/3})$$ (K.5)

However, since we are also requiring that ω be real, we see from Equations K.2 and K.3 that β^3 must be real. For $\text{Im}(\beta^3) = 0$, we must have

$$x = 1$$ (K.6)

or

$$1 - \frac{\omega^2}{\omega_{pi}^2} = \frac{\omega_{pb}^2}{\omega_{pi}^2}$$ (K.7)

This fixes the frequency at which the double root of complex β is obtained. We have now satisfied Equation K.3 and the imaginary part of Equation K.2. The remaining constraint is that the real part of Equation K.2 be satisfied. Since

$$\beta^3 = -\frac{\omega^3}{v_0^3}$$ (K.8)

for $x = 1$, we find, using Equations K.3 and K.8 in Equation K.2, that

$$\beta_{De}^2 + p^2 K_{\perp i} = -\beta_{pb}^2$$ (K.9)

The compatibility of Equations K.7 and K.9 determines the desired relation among the system parameters, which is

$$\frac{\omega_{pb}^2}{\omega_{pi}^2} = 1 - \frac{\omega_{ci}^2}{\omega_{pi}^2} - \frac{p^2}{p^2 + \beta_{De}^2 + \beta_{pb}^2}$$ (K.10)

Note, however, that Equation K.10 is not valid when $p^2 = 0$ (one-dimensional longitudinal interaction), as is clear by inspection of Equation K.9. For $p^2 = 0$, there cannot be any such transition to an absolute instability, in agreement with the analysis of Chapter 3.

We can now argue that the condition for an absolute instability is that the left-hand side of Equation K.10 be <u>larger</u> than the right-hand side (inequality given by Condition 5.15). This is a reasonable postulate because we know that there is no absolute instability when $\omega_{pb} \to 0$, so that the absolute instability should occur for ω_{pb} larger than some critical value.*

---

*This, of course, is true when the plasma waves are all forward waves, so that the right-hand side of Equation K.10 is positive (see Equation 5.16).

GLOSSARY OF COMMON SYMBOLS

$\overline{B}_0$	steady magnetic field ($\overline{B}_0 = B_0\overline{i}_z$)
c	speed of light
$\overline{E}$	electric field vector
e	charge on electron
$f_0(\overline{v})$	distribution of random velocities (subscript e or i for electrons or ions, respectively)
$\overline{H}$	magnetic field vector
$K_{\parallel}$	Equation 4.6
$K_{\perp}$	Equation 4.7
$K_{\parallel}^0$	Equation 4.4
$K_{\perp}^0$	Equation 4.5
$K_{\parallel i}$	Equation 5.3
$K_{\perp i}$	Equation 5.4
k	wave number in one-dimensional case
k_i	imaginary part of k (Im k)
k_r	real part of k (Re k)
M	ion mass
m	electron mass
$n_{p(e,i)}$	unperturbed plasma density (electrons or ions)
n_b	unperturbed beam density
p, q	radial wave numbers in a finite system

175

T_e, T_i temperature of plasma electrons and ions, respectively

u_α Alfvén speed in plasma

V_{Te}, V_{Ti} average thermal speed of plasma electrons and ions, respectively

V_0 beam voltage ($mv_0^2/2e$)

v_0 unperturbed beam velocity

β wave number of guided waves

β_{ce} ω_{ce}/v_0

β_{De} ω_{pe}/V_{Te}

β_e ω/v_0

β_i imaginary part of β (Im β)

β_{pb} ω_{pb}/v_0

β_r real part of β (Re β)

γ_0 $(1 - v_0^2/c^2)^{-\frac{1}{2}}$

η $(n_b/n_p)(T_e/2V_0)$ parameter used in Section 3.5

φ quasi-static potential ($\overline{E} = -\overline{\nabla}\varphi$)

ρ_{0b} unperturbed beam charge density (en_b)

ω radian frequency

ω_{ce} electron cyclotron frequency

ω_{ci} ion cyclotron frequency

ω_d $\omega - \beta v_0$ Doppler-shifted frequency

ω_{pb} plasma frequency of beam

ω_{pe} electron plasma frequency of plasma

ω_{pi} ion plasma frequency

ω_{p0} $(\omega_{pe}^2 + \omega_{pi}^2)^{\frac{1}{2}}$

ω_r real part of ω (Re ω)

ω_i imaginary part of ω (Im ω)

REFERENCES

1. L. D. Smullin and P. Chorney, "Propagation in Ion-Loaded Waveguides," Proc. Symp. on Electronic Waveguides, Brooklyn Polytechnic Press, Brooklyn, N.Y., 1958, pp. 229-247.

2. L. D. Smullin, "Electron-Stimulated Ion Oscillations," Notes on Plasma Dynamics, Summer Session Class Notes, M.I.T. 1959, Sec. 5.

3. R. W. Gould and A. W. Trivelpiece, "A New Mode of Wave Propagation on Electron Beams," Proc. Symp. on Electronic Waveguides, Brooklyn Polytechnic Press, Brooklyn, 1958, pp. 215-228.

4. I. Langmuir, "Scattering of Electrons in Ionized Gases," Phys. Rev., 26, 585 (1925).

5. J. R. Pierce, "Possible Fluctuations in Electron Streams due to Ions," J. Appl. Phys., 19, 231 (1948).

6. A. V. Haeff, "Space-Charge Wave Amplification Effects," Phys. Rev., 74, 1532 (1948); and "On the Origins of Solar Noise," Phys. Rev., 75, 1546 (1949).

7. A. V. Haeff, "The Electron Wave Tube: A Novel Method of Generation and Amplification of Microwave Energy," Proc. IRE, 37, 4 (1949).
   J. R. Pierce and W. B. Hebenstreit, "A New Type of High-Frequency Amplifier," Bell System Tech. J., 28, 33-51 (1949);
   L. S. Nergaard, "Analysis of a Simple Model of a Two-Beam Growing-Wave Tube," RCA Rev., 9, 585-601 (1948).

8. V. A. Bailey, "Plane Waves in an Ionized Gas with Static Electric and Magnetic Fields Present," Australian J. Sci. Research, Sec. A, 1, 351 (1948).

9. R. Q. Twiss, "On Bailey's Theory of Amplified Circularly Polarized Waves in an Ionized Medium," Phys. Rev., 84, 448-457 (1951).

10. D. Bohm and E. P. Gross, "Theory of Plasma Oscillations, Part A: Origin of Medium-Like Behavior," Phys. Rev., 75, (1949); Part B: "Excitation and Damping of Oscillations," Phys. Rev., 75, 1864 (1949).

11. A. I. Akhiezer and Ya. B. Fainberg, Dok. Akad. Nauk SSR, 64, 555 (1949); Soviet Phys. — JETP, 21, 1262 (1951).

References

12. Ya. B. Fainberg, "The Interaction of Charged Particle Beams with Plasma," J. Nucl. Energy, Part C, 4, 203 (1962).

13. F. W. Crawford and G. S. Kino, "Oscillations and Noise in Low-Pressure D. C. Discharges," Proc. IRE, 49, 1767-1788 (1961).

14. M. Sumi, "Theory of Spatially Growing Waves," J. Phys. Soc. Japan, 14, 653 (1959).

15. G. D. Boyd, L. M. Field, and R. W. Gould, "Excitation of Plasma Oscillations and Growing Plasma Waves," Phys. Rev., 109, 1393 (1958); see also Proc. Symp. on Electronic Waveguides, Brooklyn Polytechnic Press, Brooklyn, N.Y., 1958, pp. 367-375.

16. A. A. Rukhadze, "Electromagnetic Waves in Interpenetrating Plasmas," Sov. Phys.— Tech. Phys., 6, 900-905 (1962).

17. A. B. Kitsenko and K. N. Stepanov, "Excitation of Magneto-acoustic Waves in a Rarified Plasma by a Charged-Particle Beam," Sov. Phys. — Tech. Phys., 7, 215-218 (1962).

18. C. K. Birdsall, "Single-Streaming and Contra-Streaming Instabilities in a Resistive Background," U.C.R.L. No. 6631, University of California, Lawrence Radiation Lab., Livermore, Calif., May 1962.

19. L. J. Chu, "A Kinetic Power Theorem," IRE-PGED Electron Tube Research Conference, Durham, N.H., June 1951.

20. C. K. Birdsall, G. R. Brewer, and A. V. Haeff, "The Resistive-Wall Amplifier," Proc. IRE, 41, 865-875 (1953).

21. M. S. Kovner, "Instability of Low Frequency Electromagnetic Waves in a Plasma Traversed by a Beam of Charged Particles," Soviet Phys. — JETP, 13, 369-374 (1961).

22. A. I. Akhiezer, A. B. Kitsenko, and K. N. Stepanov, "Interaction of Charged-Particle Beams with Low-Frequency Plasma Oscillations," Soviet Phys. — JETP, 13, 1311-1313 (1961).

23. J. Neufeld and H. Wright, "Instabilities in a Plasma-Beam System Immersed in a Magnetic Field," Phys. Rev., 129, 1489-1507 (1963).

24. K. N. Stepanov and A. B. Kitsenko, "Excitation of Electromagnetic Waves in a Magnetoactive Plasma by a Beam of Charged Particles," Soviet Phys. — Tech. Phys., 6, 120-126 (1961).

25. F. Shimabukuro, "Unstable Modes of Drifting Charged Particles in a Plasma in a Magnetic Field," Symp. on Electromagnetic Theory and Antennas, Copenhagen, June 25, 1962.

26. T. H. Stix, The Theory of Plasma Waves, McGraw-Hill Book Company, Inc., New York, 1962.

27. E. V. Bogdanov, J. J. Kislov, and Z. S. Tchernov, "Interaction Between an Electron Stream and Plasma," Proc. Symp. on Millimeter Waves, Brooklyn Polytechnic Press, Brooklyn, N.Y., 1959, pp. 57-71.

28. M. T. Vlaardingerbroek, K. R. U. Weimer, and H. J. C. A. Nunnink, "On Wave Propagation in Beam-Plasma Systems, Phillips Res. Rep., 17, 344-362 (1962).

29. A. W. Trivelpiece, "Slow Wave Propagation in Plasma Waveguides," Tech. Rep., 7, California Institute of Technology, Pasadena, Calif. (May 1958).

30. M. A. Allen, G. S. Kino, J. Spalter, and H. Stover, "Electron Beam-Plasma Interactions," 4th International Congress on Microwave Tubes, Scheveningen, Holland, September 1962; also Internal Memo M.L. No. 958, Microwave Laboratory, Stanford University, Stanford, Calif. (October 1962).

31. M. A. Allen and G. S. Kino, "Beam Plasma Amplifiers," M. L. Report No. 833, Microwave Laboratory, Stanford University, Stanford, Calif. (July 1961).

32. M. A. Allen and G. S. Kino, "Interaction of an Electron Beam with a Fully Ionized Plasma," Phys. Rev. Letters, 6, 163 (1961).

33. M. Chodorow, J. Eidson, and G. S. Kino, "Normal Mode Theory for Electron Beam Plasma Amplification," 4th International Congress on Microwave Tubes, Scheveningen, Holland, September 1962.

34. V. J. Kislov and E. V. Bogdanov, "Interaction between Slow Plasma Waves and an Electron Stream," Proc. Symp. on Electromagnetics and Fluid Dynamics of Gaseous Plasma, Polytechnic Press, Brooklyn, N.Y., 1961, pp. 249-268.

35. P. Chorney, "Electron-Stimulated Ion Oscillations," Technical Report No. 277, Research Laboratory of Electronics, M.I.T., Cambridge, Mass. (May 26, 1958).

36. D. L. Morse, "Plasma Heating by the Electron Beam-Plasma Interaction," S.M. Thesis, Department Electrical Engineering, M.I.T., Cambridge, Mass. (1961).

37. W. D. Getty, "Investigation of Electron-Beam Interaction with a Beam-Generated Plasma," Technical Report No. 407, Research Laboratory of Electronics, M.I.T., Cambridge, Mass. (January 1963).

38. G. S. Kino and R. Gerchberg, "Transverse Field Interactions of a Beam and Plasma," Phys. Rev. Letters, 11, 185-187 (1963).

39. P. Serafim, "Analysis of Electron Beam-Plasma System," Ph.D. Thesis, Department of Electrical Engineering, M.I.T., Cambridge, Mass. (1964).

40. M. T. Vlaardingerbroek and K. R. U. Weimer, "On Wave Propagation in Beam-Plasma Systems in a Finite Magnetic Field," Phillips Res. Repts., 18, 95-108 (1963).

41. E. R. Harrison, "Stability of Longitudinal Electrostatic Oscillations in Plasmas of Finite Dimensions," Proc. Phys. Soc. (London), 79, 317-325 (1962).

42. P. Chorney and J. Valun, "Study and Investigation of the Behavior of Beam Plasma Amplifiers," Fourth Quarterly Progress Report, Microwave Associates, Inc., Burlington, Mass. (April 1961).

43. R. Q. Twiss, "On Oscillations in Electron Streams," Proc. Phys. Soc. (London), B64, 654-669 (1951).

44. R. Q. Twiss, "Propagation in Electron-Ion Streams," Phys. Rev., 88, 1392-1407 (1952).

45. L. D. Landau and E. M. Liftshitz, Electrodynamics of Continuous Media, G.I.T.T.L., Moscow, 1953, p. 141 (in Russian).

46. P. A. Sturrock, "Kinematics of Growing Waves," Phys. Rev., 112, 1488-1503 (1958). Also in Plasma Physics, edited by J. E. Drummond, McGraw-Hill Book Company, Inc., New York, 1961.

47. J. R. Pierce, Traveling-Wave Tubes, D. Van Nostrand Co., Inc., New York, 1950.

48. J. R. Pierce, "Coupling of Modes of Propagation," J. Appl. Phys., 25, 179 (1954).

49. H. A. Haus, "Electron Beam Waves in Microwave Tubes," Proc. Symp. on Electronic Waveguides, Brooklyn Polytechnic Press, Brooklyn, N.Y., 1958, p. 89.

50. H. A. Haus and D. L. Bobroff, "Small Signal Power Theorem for Electron Beams," J. Appl. Phys., 28, 694-704, 1957.

51. Ya. B. Fainberg, V. I. Kurilko, and V. D. Shapiro, "Instabilities in the Interactions of Charged Particle Beams with Plasma," Sov. Phys. — Tech. Phys., 6, 459-463 (1961).

52. R. V. Polovin, "Criteria for Instability and Gain," Sov. Phys. — Tech. Phys., 6, 889-895 (1962).

53. O. Buneman, "How to Distinguish Amplifying and Evanescent Waves," Plasma Physics, edited by J. E. Drummond, McGraw-Hill Book Company, Inc., New York, 1961.

54. P. M. Morse and H. Feshback, Methods of Theoretical Physics, McGraw-Hill Book Company, Inc., New York, 1953.

55. N. G. Van Kampen, "On the Theory of Stationary Waves in Plasma," Physica, 21, 949 (1955).

56. H. Derfler, "Theory of R. F. Probe Measurements in a Fully Ionized Plasma," Technical Report No. 106-1, Electron Devices Lab., Stanford Electronics Laboratory, Stanford University, Stanford, Calif. (June 1962).

57. A. Bers and R. J. Briggs, "Criteria for Determining Absolute Instabilities and Distinguishing Between Amplifying and Evanescent Waves," Quarterly Progress Report No. 71, Research Laboratory of Electronics, M.I.T., Cambridge, Mass. (Oct. 15, 1963) pp. 122-130.

58. H. R. Johnson, "Backward-Wave Oscillators," Proc. IRE, 43, 684-697 (1955).

59. W. H. Louisell, Coupled Mode and Parametric Electronics, John Wiley & Sons, Inc., New York, 1960.

60. L. Landau, "On the Vibrations of the Electronic Plasma," J. Phys. (Moscow), USSR, 10, 25-34 (1946).

61. W. P. Allis, S. J. Buchsbaum, and A. Bers, Waves in Anisotropic Plasmas, The M.I.T. Press, Cambridge, Mass. 1963.

62. L. R. Walker (unpublished); see J. R. Pierce, "The Wave Picture of Microwave Tubes," Bell System Tech. J., 33, 1343-1372 (1954).

63. K. Imre, "Oscillations in a Relativistic Plasma," Phys. Fluids 5, 459-466 (1962).

64. A. W. Trivelpiece and R. W. Gould, "Space Charge Waves in Cylindrical Plasma Columns," J. Appl. Phys., 30, 1784 (1959).

65. R. J. Briggs, "On the Quasi-static Analysis of Guided Waves in Plasmas and Electron Beams," Research Laboratory of Electronics Internal Memo., M.I.T., Cambridge, Mass. (October 1962) (unpublished).

66. A. E. Siegman, "Waves on a Filamentary Electron Beam in a Transverse Field Slow-Wave Circuit," J. Appl. Phys., 31, 17 (1960).

References   183

67. G. F. Freire, "Interaction Effects between a Plasma and a
    Velocity-Modulated Electron Beam," M. L. Report No. 890,
    Microwave Laboratory, W. W. Hansen Laboratory of Phys-
    ics, Stanford University, Stanford, Calif. (February 1962).

68. B. J. Maxum, "Cyclotron Wave Instabilities," Ph. D. Thesis,
    Department of Electrical Engeneering, University of
    California, Berkeley, Calif., 1963.

69. J. R. Pierce, "Limiting Stable Current in Electron Beams in
    the Presence of Ions," J. Appl. Phys., 15, 721 (1944).

70. G. C. Alexander, "Space-Charge-Wave Analysis of Dense
    Electron Beams," M. S. Thesis, Department of Electrical
    Engineering, M. I. T., Cambridge, Mass., September 1960.

71. S. Puri, "Electron Beam Interaction with Ions in a Warm-
    Electron Plasma," M. S. Thesis, Department of Electrical
    Engineering, M. I. T., Cambridge, Mass., June 1964.

72. B. R. Kusse, "Plasma Dispersion Relations and the Stability
    Criteria," M. S. Thesis, Department of Electrical Engineer-
    ing, M. I. T., Cambridge, Mass., September 1964.

73. J. M. Dawson, "Plasma Oscillations of a Large Number of
    Electron Beams," Phys. Rev., 118, 381 (1960).

74. P. A. Sturrock, "In What Sense do Slow Waves Carry Negative
    Energy?," J. Appl. Phys., 31, 2052-2056 (1960).

75. J. R. Pierce, "Momentum and Energy of Waves," J. Appl.
    Phys., 32, 2580 (1961).

76. R. J. Briggs, "Transformation of Small Signal Energy and Mo-
    mentum of Waves," J. Appl. Phys., 35 (forthcoming 1964).

77. R. M. Fano, L. J. Chu, and R. B. Adler, Electromagnetic
    Fields, Energy, and Forces, John Wiley & Sons, Inc.,
    New York, 1960.

INDEX